URBAN CLIMATE MITIGATION TECHNIQUES

The urban climate is continuously deteriorating. Urban heat lowers the quality of urban life, increases energy needs, and affects the urban socio-economy. *Urban Climate Mitigation Techniques* presents steps that can be taken to mitigate this situation through a series of innovative technologies and examples of best practices for the improvement of the urban climate.

Including tools for evaluation and a comparative analysis, this book addresses anthropogenic heat, green areas, cool materials and pavements, outdoor shading structures, evaporative cooling and earth cooling. Case studies demonstrate the success and applicability of these measures in various cities throughout the world.

Useful for urban designers, architects and planners, *Urban Climate Mitigation Techniques* is a step by step tour of the innovative technologies improving our urban climate, providing a holistic approach supported by well-established quantitative examples.

Mat Santamouris is Professor of Energy Physics at the University of Athens, Greece, and Professor of High Performance Architecture at the University of New South Wales, Australia. In addition to many international appointments as visiting professor, he is the Editor-in-Chief of *Energy and Buildings*, and Associate Editor of the *Solar Energy Journal*. Professor Santamouris is the editor of the Routledge series *Buildings Energy and Solar Technology* and the editor and author of 12 international books on topics related to heat islands, solar energy, and energy conservation in buildings.

Denia Kolokotsa is Assistant Professor at the Technical University of Crete; Lab Director of Energy Management at the Built Environment Research Lab; Board Director of European Cool Roofs Council; Editor-in-Chief of *Advances in Building Energy Research*; and Editorial Board Member of *Energy and Buildings and Renewable Energy*. Her research interests include energy management and energy efficiency in the built and urban environment, neural networks and fuzzy logic technology, advanced optimization theory and applications, and decision support systems in energy modelling and automation.

URBAN CLIMATE MITIGATION TECHNIQUES

Edited by
Mat Santamouris and
Denia Kolokotsa

LONDON AND NEW YORK

First published 2016
by Routledge
2 Park Square, Milton Park, Abingdon, Oxon OX14 4RN

and by Routledge
52 Vanderbilt Avenue, New York, NY 10017

First issued in paperback 2020

Routledge is an imprint of the Taylor & Francis Group, an informa business

British Library Cataloguing-in-Publication Data
A catalogue record for this book is available from the British Library

Library of Congress Cataloging-in-Publication Data
Urban climate mitigation techniques / edited by Mat Santamouris and
    Denia Kolokotsa.
    pages cm
    Includes bibliographical references.
    1. Climate change mitigation. 2. Urban climatology. 3. City planning.
    I. Santamouris, M. (Matheos), 1956– editor. II. Kolokotsa, Denia,
    editor.
    TD171.75.U73 2016
    363.738′747—dc23
    2015023165

ISBN 13: 978-0-367-66998-0 (pbk)
ISBN 13: 978-0-415-71213-2 (hbk)

Typeset in Corbel
by Florence Production Ltd, Stoodleigh, Devon, UK

CONTENTS

FIGURES AND TABLES

Figures

Tables

ABOUT THE EDITORS

Mat Santamouris is Professor of Energy Physics at the University of Athens, Professor of High Performance Architecture in the University of New South Wales, Australia, and serves or has served as visiting professor at London Metropolitan University, Tokyo Polytechnic University, the National University of Singapore, Bolzano University, Brunel University, London, and the Cyprus Institute.

He is editor of the series Buildings, Energy and Solar Technologies, published by Earthscan. He served as President of the National Center of Renewable Energies and Savings for the period 2010–2012.

He is Editor-in-Chief of *Energy and Buildings* and past Editor-in-Chief of the *Journal of Building Environmental Research*, Associate Editor of the *Solar Energy Journal*, Consulting Editor of the *Journal of Sustainable Energy*, member of the Editorial Advisory Board of the *Journal of Energy Conversion and Management*, and member of the Editorial Board of eight additional journals.

He is editor or author of 14 international books on topics related to heat island mitigation technologies, solar energy and energy conservation in buildings published by international publishers, and guest editor of twenty special issues published by selected scientific journals. He is the coordinator of many major international research programs, such as ZERO PLUS, PASCOOL, OFFICE, POLISTUDIES, AIOLOS, BUILT, RESET, INTERSET, COOL ROOFS, etc., and is a consultant for many international and national energy institutions.

He is an external examiner at fourteen international universities and referee in more than 100 international peer review scientific journals, and a reviewer of national and international research projects in the European Commission, USA, UK, Canada, France, Germany, Italy, Singapore, Sweden, Luxembourg, Ireland, Estonia, Korea, Belgium, Slovenia, Qatar, Cyprus, etc. He is also the author of almost 250 scientific papers published in peer-reviewed international scientific journals.

He has received many prizes and awards, including: Grand Award, Professional Green Building Council, PGBC, 2006; Hong Kong IA Research Award, 2006; National Energy Globe Award, 2006; Sustainable Energy Europe Award, 2007; Best and Outstanding Paper Award published in *Solar Energy Journal* during 2005–2006; nominated for The Sir Robert McAlpine International Book Award, London, 1997; National Award for Environmental Research, ECOCITY 2008; PROSE Award, 2012; Reference Work: Best Multivolume Reference/Science, by the American Association of Publishers, 2013; European Award on Energy Efficient Buildings, European Competition, 2014; ECOCITY Award for Best Scientific Study on Environment, 2015.

Denia Kolokotsa is currently an Assistant Professor at the School of Environmental Engineering of the Technical University of Crete, Greece. Her research interests include energy management for the built environment, energy efficiency and renewables. Moreover she has developed expertise in the field of urban dynamics and environmental and ecological issues at the metropolitan and regional scale with emphasis on the urban heat island mitigation and adaptation strategies. She is the author of more than 100 papers published in high impact scientific journals and conference proceedings. She is the Editor-in-Chief of the journal *Advances in Building Energy Research*, as well as an Editorial Board Member for *Renewable Energy and Energy and Buildings*. Finally she has participated in more than 25 European and national projects, and coordinated 3 EU (FP7 and Horizon) projects and 3 national projects.

ABOUT THE CONTRIBUTORS

Servando Alvarez Domínguez is full Professor of the Department of Energy Engineering at the University of Seville. He was in charge of the basic research (including modeling, experiments and validation) as well as the design and evaluation of the climatic control of the outdoor spaces of the EXPO '92 Universal Exhibition of Seville. He has participated in more than 50 international research projects related to urban climate and energy performance of buildings. He coordinates the transposition to Spain of the EPBD in the activities related to reduction of the heating and cooling energy needs.

Nektarios Chrysoulakis is Director of Research at the Foundation for Research and Technology – Hellas (FORTH) in Heraklion, Greece. He holds a BSc in Physics, an MSc in Environmental Physics and a PhD in Remote Sensing from the University of Athens. He has been involved in R&D projects funded by organizations such as the European Union, the European Space Agency and the Ministries of Environment, Development, Culture and Education. He is the coordinator of the H2020 (Space) project URBANFLUXES (http://urbanfluxes.eu). Dr. Chrysoulakis was also the Coordinator of the FP7 projects BRIDGE (http://www.bridge-fp7.eu), dealing with Urban Metabolism, as well as of the projects GEOURBAN (http://geourban-fp7-eranet.com) and SEN4RUS (http://sen4rus.eu), both dealing with the development and on-line evaluation of satellite-based indicators for urban planning and management. He has more than 150 publications in peer-reviewed journals and conference proceedings.

Argiro Dimoudi is Assistant Professor of Science and Technology of Structures with emphasis on Environmental Design at the Department of Environmental Engineering at Democritus University of Thrace (GR). She has extensive experience on aspects of energy conservation and RES in buildings, environmental friendly materials, microclimate and outdoor spaces, bioclimatic design, and energy consulting. She has experience as coordinator, project manager and researcher in research, demonstration and non-technological projects on energy and sustainability in buildings and cities. She has authored several books, papers in international scientific journals, conference presentations (International and National) and technical reports.

Christine Georgatou graduated in 2005 from the School of Environmental Engineering at the Technical University of Crete. Since then, she has been working in the public sector (Administration of Environmental and Rural Planning of Crete, Technological Educational Institute of Crete) and as a Freelance Engineer in the field of environmental impact assessments and implementation of RES facilities in Greece. She is also certified on energy certificates for buildings for the national EPBD implementation. In the recent years she has been participating in European and national projects relative to energy efficiency and RES. She is an experienced user of software tools (i.e Matlab, ArcGis, EnergyPlus). Her working and research interests are in the field of microscale and building energy modelling as well as time series forecasting techniques.

C.S.B. Grimmond joined the Department of Meteorology, University of Reading in August 2013. Previously she was a Professor at King's College London, after being Assistant, Associate and Full Professor at Indiana University, Bloomington USA. She is a past President of the International Association of Urban Climate (IAUC) and past Lead Expert for the WMO on Urban and Building Climatology. Sue is an Editor for *Journal of Atmospheric and Oceanic Technology*, and on the editorial boards of *Journal of Applied Meteorology and Climatology* and *Urban Climate*. In 2006 she was elected Fellow of the American Meteorological Society and awarded Doctor of Science Honoris Causa from Göteborg University, Sweden. In 2008 she was awarded the Universitatis Lodziensis Amico Medal from University of Łódź, Poland. She was the 2009 recipient of the Helmut E. Landsberg Award from the American Meteorological Society and the 2009 recipient of the Luke Howard Award from the International Association for Urban Climate. She is a Special Expert to Shanghai Institute of Meteorological Science, Shanghai Meteorological Service, China Meteorological Administration. She currently is an advisor to projects in Portugal (CLICUB), South Korea (WISE), USA (NEON), and WMO (GURME, Megacity Implementation Plan).

Maria Kaltsa holds architecture degrees from Cooper Union (BArch) and Yale University (MArch). As a practicing architect for over 25 years, she has designed and constructed many projects of diverse scales and participated in research projects as a scientific collaborator of practice. Her office was part of the team that received the First Prize in the architectural competition for the rehabilitation of the three-kilometer-long sequence of public spaces at the foot of the Acropolis; in 2004 she shared the Architect of the Year award for the finished project, which became a favorite European promenade. She recently received a First Prize in the competition for the design of a Zero Energy landmark new building for the city of Thessaloniki. From 2009–2012 she served as General Secretary of Regional Planning and Urban Development of the Greek Environment, Energy & Climate Change Ministry, where she drafted the new National Building Code in effect today, adapting green parameters. She contributed to the production of regulatory plans as frameworks for the sustainable development of metropolitan regions, initiated architectural competitions, calls for Greece's participation at the Venice Architecture Biennale, and managed the early stages of major urban interventions such as Re-Think Athens to restructure the city center, and a new metropolitan park at Faliro Bay.

Steve Kardinal Jusuf is an Assistant Professor in the Engineering Cluster, Singapore Institute of Technology. He has a PhD in Building Science from the Department of Building, National University of Singapore. He received the World Future Foundation PhD Prize in Environmental and Sustainability Research for his PhD work on air temperature prediction model within urban climatic mapping method. His research interests are in the area of urban microclimate and urban climatic mapping with Geographical Information Systems (GIS). He has worked in a number of research projects with various Singapore government agencies, mainly in the topic of urban climatic mapping for sustainable urban development.

Francisco José Sánchez de la Flor is a lecturer of the Department of Thermal Machines and Engines at the University of Cádiz. He has participated in different international research projects related to urban climate and energy performance of buildings. In these topics, he was in charge of the development of characterization models for the detailed calculation of the heat transfer processes, and their implementation in software tools. As part of the Termotecnia Research Group of the University of Seville, he has participated in the transposition to Spain of the EPBD.

Afroditi Synnefa is a Building and Environmental Physicist. She works as a Research Associate in the Group of Buildings and Environmental Studies at the National and Kapodistrian University of Athens (NKUA). She graduated from the Physics Department of NKUA and received a DEA in Buildings and Environment from the Ecole Nationale des Travaux Publics de l'Etat, in Lyon, France. Afroditi has a PhD in Environmental Science from the Physics Department of NKUA. She is a Consultant and Technical Committee Leader for the European Cool Roofs Council. She has teaching experience at the undergraduate and postgraduate levels, and in professional education and e-learning. She has a large number of publications in international scientific peer-reviewed journals, conference proceedings and books.

Nyuk Hien Wong is Professor in the Department of Building, National University of Singapore. Professor Wong was the principal investigator of a number of research projects related to urban heat islands, urban climatic mapping and greenery, in collaboration with the various government agencies, such as the National Parks Board (NParks), Housing Development Board (HDB), National Environmental Agency (NEA) and Building Construction Authority (BCA). Professor Wong has been published in more than 150 international refereed journals and conference papers and was the co-author of three books on rooftop and urban greenery. He has also been invited to serve in various advisory committees both locally and internationally.

Stamatis Zoras is Assistant Professor of Energy Design and Efficiency in the Department of Environmental Engineering at Democritus University of Thrace. He has worked as Senior Building Physicist at WYG Consulting Engineers in London and as Director of an ISO17025-certified laboratory. His current activities include environmental CFD, urban microclimate and reformations, earth-contact heat-transfer, air pollution and epidemic information dissemination to the public. He has participated or coordinated numerous national and European research projects in the field of environmental sciences and urban simulation. He has authored over 70 publications in journals, conference proceedings, books and reports.

1

URBAN WARMING AND MITIGATION

Actual status, impacts and challenges

Mat Santamouris

University of Athens – Athens, Greece

Introduction

Almost four billion people actually live in cities and official forecasts predict a very large increase in the next few years (United Nations, 2014). Urban population is increasing rapidly because of childbirth rates and the important migration of the rural population into cities caused by expectations for a better life, local conflicts and lack of resources in the country areas. As mentioned by PWC (2014), the increase of the world's urban population is close to 1.5 million people per week. High density and increased consumption patterns make cities the higher consumer of global resources as they are responsible for almost 75 per cent of the world's assets (PWC, 2014). In parallel, the urban environment experiences a very significant territorial expansion, known as urban sprawl, combined with a significant change of land use. Losses in green spaces and scaling of urban land because of the extensive use of opaque surfaces of paving, in combination with a very high increase of released anthropogenic heat, have affected the urban climate, resulted in a serious environmental degradation and have increased significantly the urban ecological footprint (Oke, 1997).

Global climatic change caused by the increased concentration of greenhouse gases in the atmosphere contributes to increased urban temperatures and the frequency and length of extreme climatic phenomena like heat waves. The magnitude of the ambient temperature increase caused by the global climate change is forecasted by the IPCC Committee. For the period 1990–2005, predictions indicated an increase between 0.15 K to 0.3 K, which is already confirmed by measurements (IPCC, 2014). At the same time, predictions for the period 1990–2100 indicate that the possible ambient temperature increase will range between 1.8 K to 4 K.

Additionally with global climate change, the thermal balance of cities is highly affected by the increased absorption of solar radiation, the corresponding increase of sensible heat released by urban structures, higher anthropogenic heat, reduced latent heat, higher emission of infrared radiation and other specific sources (Landsberg, 1981). Additional heat accumulated and released in the urban environment results in a positive thermal balance and increased urban ambient temperatures compared to the surrounding urban environment. Such a phenomenon is known as the 'urban heat island' and it is the most documented phenomenon of climate change. The difference between the urban temperature and the corresponding rural or suburban one is referred as the 'urban heat island intensity' and its magnitude is a function of the physical, structural and morphological characteristics of the cities, the urban layout, local climatic parameters, the synoptic weather conditions and also the total anthropogenic heat generated and released in the city (Oke *et al.*,1991). Studies on the heat island characteristics are available for most of the medium and large cities in the world and the reported urban heat island intensities reach values up to 10 K.

Global climatic change and urban heat islands affect the urban climate, but their specific impact may not always act in a synergistic way. The specific mechanisms of the interaction between global climatic change, in particular heat waves, and urban heat islands are analysed by Li and Bou-Zeid (2013). Three mechanisms of interrelationship are identified:

1. Given that heat waves strengthen secondary circulations, the warm air over the city moves upwards, and fresh and cool air enters the city from the neighbouring zones. This mechanism has a negative feedback on urban heat islands.
2. Heat waves are mostly associated with low wind speeds. Under the same conditions the urban heat island is always stronger. Thus, under the specific conditions the two phenomena may have a synergistic effect.
3. During heat waves, the surface temperatures are increasing and thus more evapotranspiration and storage to the ground occurs. Given that in the rural areas there is a higher availability of surface moisture, the temperature difference between the urban and rural areas tends to increase.

Simulations and future projections on the possible impact of global climatic change on the magnitude of urban heat islands are quite contradictory. According to Oke (1997), the magnitude of the urban heat island may not be modified even for more intensive conditions of climate change, while Brázdil and Budíková (1999) predicted a lower magnitude of the urban heat island under intensified global climatic change conditions because of the possible increase of the vertical instability and the corresponding dissipation of the heat in the urban environment. Another study performed by McCarthy *et al.* (2010) concluded that stronger climatic change phenomena may cause a reduction of the urban heat island magnitude of 6 per cent; however, in urban zones with an intensive urbanization, urban heat island may increase up to 30 per cent.

Increased urban temperatures have a serious impact on the global environmental quality of cities. Urban warming increases the energy consumption for cooling purposes, and the peak electricity demand during the summer period raises the concentration of harmful pollutants like the tropospheric ozone and VOCs, increases the emissions of CO_2 to the atmosphere, deteriorates indoor and outdoor thermal comfort during the warm periods, seriously affects health conditions, and increases mortality, while it has an important environmental and economic impact. A quite detailed analysis of the specific impact of urban warming on energy, environment, health and thermal comfort is presented in the following chapters.

To counterbalance the impacts of urban warming, specific mitigation and adaptation technologies are proposed, developed and are already in use in large-scale projects. In particular, the main objective of the specific mitigation technologies is to weaken the strength of heat sources and force the potential of heat sinks in the urban environment. Four major clusters of mitigation technologies are identified as the most promising ones and where most of the research takes place:

1. Mitigation technologies aiming to decrease absorption of the solar radiation in the urban environment. This is mainly achieved through the use of materials highly reflective to solar radiation presenting also a high emissivity factor. These materials, known as cool materials, are used in buildings' facades and roofs, or in urban structure, pavements, etc. Lower absorption of the solar radiation and high emissivity factor decrease the surface temperature of the urban fabric and minimizes the corresponding release of sensible heat to the atmosphere.
2. Technologies aiming to increase evapotranspiration in the urban environment. This may be achieved through the intensive use of urban greenery like urban parts and green roofs and also through the use of water-retentive pavements.
3. Technologies and systems aiming to dissipate the excess urban heat into a sink of much lower temperature like the ground.
4. Techniques and technologies aiming to decrease the release of anthropogenic-generated heat in the urban atmosphere.

Urban heat island: Existing studies and knowledge

The term 'urban heat island' is used to describe either the increase of the ambient temperature of a city compared to the surrounding rural zones, or the increase of the urban surface temperature. The present book deals with the so-called 'near surface urban canopy layer heat island', which signifies the increase of the urban ambient temperature measured at 2 meters height against the corresponding ambient temperature in the rural areas. As already mentioned the urban heat island is the most documented phenomenon of climate change. Measurements of the urban–rural temperature differences are available since the end of the 19th century and experimental data on the characteristics of the urban heat island are now available for more than 200 cities around the world. According to Santamouris (2015a), experimental data are available for 225 cities in the world and in particular for 101 cities and regions in Asia and Australia, 72 cities in Europe, 43 cities in the Americas and 9 cities in Africa. Organised information on the magnitude of the urban heat island in Asia and Australia is given in Santamouris (2015b) and for Europe in Santamouris (2007). Table 1.1 gives specific information on the heat island studies in Europe as well as information on the heat island intensity of the corresponding cities (Santamouris, 2007). Also, Table 1.2, summarises the existing urban heat island experimental studies in Asia and Australia.

The development of the urban heat island is determined by the regional and local characteristics as well as from the specific weather conditions. It is well known that the phenomenon is better developed under anticyclonic synoptic weather conditions while its intensity is seriously reduced under cyclonic conditions (Mihalakakou et al., 2002). Local weather conditions have a strong impact on the development of an urban heat island (UHI). Higher urban–rural differences are better developed under low wind speeds and clear sky conditions. It seems that the impact of wind speed is more important than that of the cloud cover as important UHI intensities are reported under relatively cloudy conditions combined with very low wind speeds (Kim & Baik, 2004). The maximum wind speed over which the UHI tends to be weaker is a function of the characteristics of the city and in particular of its size. Threshold values of wind speed vary between 4 to 11 m/sec (Park, 1986).

Table 1.1 Characteristics of Urban Heat Island Studies in Europe (Santamouris, 2007)

City	Type of Measurements	Period of Measurements	Maximum Heat Island Intensity	Time of Occurrence
Southern Europe				
Athens, Greece	Almost 30 surface stations in and around the city. Also, a mobile station	1996–2006	12°C	Daytime, 1–3 p.m.
Rome, Italy	Ten urban and rural surface stations	1964–1975	4.3°C in min. temperatures	Not reported, but apparently at night
Parma, Italy	One urban and one rural surface station	1959–1973	1.6°C	During daily max. temperature
Florence, Italy	Surface and mobile stations	Summer 2005	3.0°C during summer	2–6 p.m.
Milan, Italy	One urban station and the airport surface station	Before 1992	1.4°C	Not reported
Lisbon, Portugal	Mobile station	4 and 15 January 1995	3.5°C	Nighttime
Oporto, Portugal	Mobile station	November 2003 to January 2005	7.3°C	
Aveiro, Portugal	Mobile station	48 nights in the summer, autumn and winter of 1996	7.5°C	Hours after sunset
Madrid, Spain	Four surface stations	1965–1987	3.1°C in min. temperatures	Not reported, but apparently at night
Granada, Spain	Mobile and surface stations	Not reported	5.0°C	Early morning
Izmir, Adana, Bursa and Gaziantep, Turkey	One urban and rural surface station for each city	1951–1990	6.5–9.0°C	Hours after sunset
Istanbul, Turkey	Seven urban and rural surface stations	1951–1992	2°C in min. temperatures	Not reported, but apparently at night
Central Europe and UK				
Bucharest, Romania	Three ground surface stations	May–December 1994	3.5°C	Hours after sunset
Szeged, Hungary	Mobile and stationary stations	March and February 2000 and April–October 2002	2.1°C in min. temperatures during the heating season, and 3.1°C during the non-heating season	Not reported, but apparently at night
Debrecen, Hungary	Mobile station	April 2002 and March 2003	5.8°C	Hours after sunset
Warsaw, Poland	Surface air stations	June 2001	1.1°C mean daily temp. 3.1°C min. temp.	
Poznan, Poland	Surface air stations at the airport and center of the city	14–30 June 2005	0.1°C max. temp. 2.1°C mean temp. 2.3°C min. temp.	3.8°C max. temp.

Table 1.1 *continued*

City	Type of Measurements	Period of Measurements	Maximum Heat Island Intensity	Time of Occurrence
Bydgoszcz, Poland	One urban and one rural station	17 July to 30 September 2004	1.1°C mean temp. 1.2°C min. temp. 1.5°C max. temp.	
Glucholazy, Poland	One urban and one rural station	14–26 June 2001	0.8°C mean temp. 1.9°C min. temp. 3.5°C max. temp.	
Wroclaw, Poland	Six surface urban and rural stations and mobile stations	1997–2000	8–9°C 5–6°C	Nighttime Daytime
Lodz, Poland	One urban and one rural station	Since 1992	12°C	The greatest differences occur during summer nights when skies are clear.
Prague, Czech Republic	An urban station located in the city centre and three rural stations	1961–1990	Annual trend is close to 1.2°C per 100 years, while the corresponding trend for summer is close to 1.5°C/100 years. Analysis is based on daily minimums	Not reported, but apparently at night
Paris, France	One urban site and one rural site have been instrumented with sodars, lidars and surface measurements. Additional radiosondes, 100 m masts and Eiffel Tower data were also collected	Winter 1995	6.0°C	8 a.m.
Basel, Switzerland	Seven energy balance stations	2001–2002	3.0°C	After sunset
Vienna, Austria	Six urban and three rural surface temperature stations		1.6°C	
Gratz, Austria	Surface stations	1995	4.3°C, based on min. temperatures.	Nighttime
London, UK	68 surface stations	Summer 1999	7.0°C	Nighttime
Northern Europe				
Moscow, Russia	Eight urban and rural surface stations	1990	9.8°C	Nighttime
Gotemborg, Sweden	Urban surface stations and mobile stations	1988–1991	6.0°C	Nighttime
Oulu, Finland	Three surface stations	1996–1998	3.4°C	Winter period

Table 1.2 Characteristics of the Heat Island Studies in Asia and Australia (Santamouris 2015b)

No.	City /Country	Period of Measurements	Type of Measurements	Duration of Measurements	Number of Stations	Urban Heat Island Intensity
1	Adelaide, Australia	2000–2001	Non-Standard Fixed Measuring Stations	11 months	4 stations	Max. Daytime (–1.6)–(–3.8) Max. Nighttime 6–8°C
2	Melbourne, Australia	1972–1991	Standard Fixed Measuring Stations	20 years	4 stations	Mean Annual Nighttime: 1.13°C Mean Max.: 2.68°C
2b	Melbourne, Australia	1973–1991	Standard Fixed Measuring Stations	19 years	4 stations	Max. Annual Nighttime: 6.0°C Mean Annual Max.: 2.68°C
2c	Melbourne Australia	1992	Mobile Stations	Several days	Several stops	Max. during transect: 7.1°C (3 hours after sunset)
3	Camperdown, Australia	1994	Mobile Stations	Several days	Several stops	Max. during transect: 2.7°C (after sunset)
4	Colac, Australia	1994	Mobile Stations	Several days	Several stops	Max. during transect: 2.8°C (nighttime)
5	Hamilton, Australia	1994	Mobile Stations	Several days	Several stops	Max. during transect: 5.4°C (nighttime)
6	Hobart, Australia	1979	Mobile Stations	Not Known		Max.: 5°C
7	Sydney, Australia	1965–2007	Standard Fixed Measuring Stations	42 years	2 stations	Difference of mean max. temperature in January: 3.5°C
8	Christchurch, New Zealand	1979	Mobile Stations	1 day	120 places	Nocturnal Average during winter: 5–11°C (early afternoon and evening)
8a	Christchurch, New Zealand	1967–1968	Mobile Stations	Several days in 7 months	Several places; stops	Nocturnal Average: 6.7°C
9	Seoul, Korea	2001	Standard Fixed Measuring Stations	12 months	Not known	Average Annual 2.0°C
9b	Seoul, Korea	1973–1996	Standard Fixed Measuring Stations	23 years	2 stations	Average Max. Annual: 2.3–3.4°C Absolute Max.: 4.5 K
9c	Seoul, Korea	2001–2002	Standard Fixed Measuring Stations	1 year	31 stations	Average Max.: 2.2 k ranging 0.6–3.4°C Max. Max.: 8°C
9d	Seoul, Korea	1999–2002	Standard Fixed Measuring Stations	4 years	11 stations	Average Daily Max.: 3.5–4.8°C Daily Max.: 8°C
9e	Seoul, Korea	1970–2001	Standard Fixed Measuring Stations	31 years	2 stations	Average Annual Max.: 3.3°C Annual Max.: 2.8– 4.0°C Average Daily Max.: 4.8 K
9f	Seoul, Korea	1982	Mobile Traverses	June to August	Up to 50 observation points	Max. Intensity: 7 K
10	Incheon, Korea	1970–2001	Standard Fixed Measuring Stations	31 years	2 stations	Average Annual Max.: 2.48°C Annual Max.: 2.0–3.3°C

Table 1.2 *continued*

No.	City /Country	Period of Measurements	Type of Measurements	Duration of Measurements	Number of Stations	Urban Heat Island Intensity
11	Daejeon, Korea	1970–2001	Standard Fixed Measuring Stations	31 years	2 stations	Average Annual Max.: 2.34°C Annual Max.: 1.5–3.5°C
12	Daegu, Korea	1970–2001	Standard Fixed Measuring Stations	31 years	2 stations	Average Annual Max.: 3.1°C Annual Max.: 2.2–4.0°C
13	Gwanglu, Korea	1970–2001	Standard Fixed Measuring Stations	31 years	2 stations	Average Annual Max.: 2.59°C
14	Busan, Korea	1970–2001	Standard Fixed Measuring Stations	31 years	2 stations	Average Annual Max.: 2.2°C
15	Tokyo, Japan	2001	Standard Fixed Measuring Stations	12 months	Not known	Average Annual 2.0°C
15a	Tokyo, Japan	1992	Mobile Stations	One day	359 points	Nighttime: 8°C
15b	Tokyo, Japan	1994–1995	Standard Fixed Measuring Stations	One year	3 stations	*Winter*: Max. Daytime: 3.2°C Min. Daytime: 0.2°C Max. Night time: 3.0°C Min. Nightime: 2.7°C *Summer*: Max. Daytime: 1.1°C Min. Daytime: -1.5 °C Max. Night time: 1.5°C Min. Nightime: 1.0°C
16	Osaka, Japan	2001	Standard Fixed Measuring Stations	12 months	Not known	Average Annual Max.: 3.1°C
16b	Osaka, Japan	2005	Mobile Stations	5 months	Various stops in arious parts of the city. No rural station	Daily Max.: 5°C
16c	Osaka, Japan	2005	Standard Fixed Measuring Stations	3 months	24 stations in various parts of the city. No rural station	Average Max.: 1–3°C
17	Kumamoto, Japan	1980	Mobile Stations	5 days	450–500 points	Max. Daytime: 3°C Max. Nightime: 2°C
18	Kyoto, Japan	2002	Mobile Stations	7 days	6 points	Max. Daytime: 2.1°C
19	Tachikawa City, Tokyo area, Japan	1983	Mobile Stations	5 months	Several points	Daytime: 3.5°C
20	Fuchu City, Tokyo area, Japan	1983	Mobile Stations	5 months	Several points	Daytime: 1.2°C
21	Fussa City, Tokyo area, Japan	1983	Mobile Stations	5 months	Several points	Daytime: 1.0°C
22	Higashimurayama City, Tokyo area, Japan	1983	Mobile Stations	5 months	Several points	Daytime: 0.4°C
23	Akigawa City, Tokyo area, Japan	1983	Mobile Stations	5 months	Several points	Daytime: 1.0°C
24	Sedai, Japan	2001–2005	Standard Fixed Measuring Stations	60 years data	2 stations	Average annual: 1.5°C

Table 1.2 *continued*

No.	City /Country	Period of Measurements	Type of Measurements	Duration of Measurements	Number of Stations	Urban Heat Island Intensity
24b	Sedai, Japan	2007–2008	Standard Fixed Measuring Stations	1 year	5 stations	Daily Variation: (−1°C)–*(5.0°C)
25	Nagano Area, Japan	2001–2002	Mobile Stations	90 days	Several points	Night Max.: 8°C
26	Kumagaya City, Japan	2010–2012	Standard Fixed Measuring Stations	3 years	2 stations	Average Temp: 0.17–0.90°C Max. Temp: 0.55–1.63°C Min. Temp: 0.2–0.62°C
27	Hakuba, Japan	2000–2001	Mobile Stations	2 years	Several stops	Max.: 4.1°C
28	Matsumoto City, Japan	1999–2001	Mobile Stations	2 years	Several stops	Max.: 6.1°C
29	Asashina Japan	1998–1999	Mobile Stations	2 years	Several stops	Max.: 2.7°C
30	Akaiwa and Tokida, Japan	1998–1999	Mobile Stations	2 years	Several stops	Max.: 2.6°C
31	Tomono, Japan	1998–1999	Mobile Stations	2 years	Several stops	Max.: 1.7°C
32	Koundai, Japan	1998–1999	Mobile Stations	2 years	Several stops	Max.: 3.2°C
33	Obuse, Japan	1996–1997	Mobile Stations	153 times	Several stops	Max.: 5.4°C
34	Kofu Basin, Japan	2008–2010	Mobile Stations	3 summers	38 points	Nighttime: 2°C
35	Tsukuba City, Japan	2008	Mobile Stations	3 days	25 points	2.5°C
36	Taipei, Taiwan	2001	Standard Fixed Measuring Stations	12 months	Not known	Average Annual 2.0°C
36b	Taipei, Taiwan	2006	Standard Fixed Measuring Stations and Other Equipment	Several days	10 stations	Max. Daytime: 4°C Max. Nightime: 4–6°C
37	Manila, Philippines	2001	Standard Fixed Measuring Stations	12 months	Not known	Average Annual 1.4°C
38	Bangkok, Thailand	2001	Standard Fixed Measuring Stations	12 months	Not known	Average Annual. 0.5°C
38b	Bangkok, Thailand	2004–2008	Standard Fixed Measuring Stations	4 years	2 stations	Mean Max.: 2.4°C Mean Min.: 0.42°C
39	Chiang Mai, Thailand	2004–2008	Standard Fixed Measuring Stations	4 years	2 stations	Mean Max.: 2.7°C Mean Min.: 0.77°C
40	Songkhla, Thailand	2004–2008	Standard Fixed Measuring Stations	4 years	2 stations	Mean Max.: 2.4°C Mean Min.: 0.53°C
41	Jakarta, Indonesia	2001	Standard Fixed Measuring Stations	12 months	Not known	Average Annual 0.5°C
42	Bahrain	2009	Standard Fixed and Mobile Measuring Stations	2 summer days	14 stations	Average daytime 2–5°C
43	Kuwait	1958–1980	Standard Fixed Measuring Stations	22 years	3 stations	Average Annual: 0.84°C
44	Muscat, Oman	2007–2008	Mobile and Fixed Standard Measuring Stations	2 months	14–20 points	Summer:0.0–4.3°C Winter: (−0.2)– (2.0) Absolute Max.: 6.2°C
45	Beijing, China	2005	Standard Fixed Measuring Stations	One month	60	Average Max. Daytime 1.2°C Absolute Daytime Max.: 5.3°C Absolute Daytime Min.: −1.6°C Max. Nightime: 1.62°C

Table 1.2 *continued*

No.	City /Country	Period of Measurements	Type of Measurements	Duration of Measurements	Number of Stations	Urban Heat Island Intensity
45b	Beijing, China	1961–2000	Standard Fixed Measuring Stations	39 years	7	Annual Average of Max. Temp: 0.5°C Annual Average of Min. Temp.: 0.0°C
45c	Beijing, China	1977–2000	Standard Fixed Measuring Stations	23 years	2	Annual Max. Average Temperatures: 2°C Annual Max. Min. Temp.: 2.13°C
45d	Beijing, China	1993–2003	Standard Fixed Measuring Stations	11 years	11 stations	Average Annual: 0.1–2°C
46	Wuhan, China	1961–2000	Standard Fixed Measuring Stations	39 years	5	Annual Average of Max. Temp.: 0.3°C Annual Average of Min. Temp: 0.0°C
47	North East China	1954–1983	Standard Fixed Measuring Stations	29 years	One rural and one urban per city	Annual Average: 0.04°C Spring Average: 0.18°C Summer Average: 0.03°C Autumn Average: 0.09°C Winter Average: 0.06°C
48	Northern Plains, China	1954–1983	Standard Fixed Measuring Stations	29 years	One rural and one urban per city	Annual Average: 0.63°C Spring Average: 0.18°C Summer Average: 0.20°C Autumn Average: 1.07°C Winter Average: 1.42°C
49	Middle Lower China	1954–1983	Standard Fixed Measuring Stations	29 years	One rural and one urban per city	Annual Average: 0.36°C Spring Average: 0.44°C Summer Average:0.42°C Autumn Average:0.38°C Winter Average: 0.21°C
50	South East Coast, China	1954–1983	Standard Fixed Measuring Stations	29 years	One rural and one urban per city	Annual Average: 0.01°C Spring Average: -0.13°C Summer Average:0.20°C Autumn Average:0.03°C Winter Average: −0.08°C
51	South West China	1954–1983	Standard Fixed Measuring Stations	29 years	One rural and one urban per city	Annual Average: 0.00°C Spring Average: 0.24°C Summer Average:0.24°C Autumn Average:−0.05°C Winter Average: −0.22°C
52	North West China	1954–1983	Standard Fixed Measuring Stations	29 years	One rural and one urban per city	Annual Average: 0.35°C Spring Average: 0.32°C Summer Average:0.35°C Autumn Average: 0.36°C Winter Average: 0.35°C
53	Shanghai, China	1959–2005	Standard Fixed Measuring Stations	46 years	11 stations	Mean Annual: 0.41°C in the 60s to 1.44°C in the 90s to about 2.0°C in 2005
53b	Shanghai, China	1975–2004	Standard Fixed Measuring Stations	30 years	10 stations	Mean Annual: 0.7°C
53c	Shanghai, China	1975–2004	Standard Fixed Measuring Stations	30 years	11 stations	Mean Annual: 0.7–1.3°C

Table 1.2 *continued*

No.	City /Country	Period of Measurements	Type of Measurements	Duration of Measurements	Number of Stations	Urban Heat Island Intensity
53d	Shanghai, China	1961–1997	Standard Fixed Measuring Stations	37 years	16 stations	Mean Annual: 0.6–1.0°C Absolute Max.: 5°C
54	Large Size Cities, China	1961–2000	Standard Fixed Measuring Stations	39 years	60 urban stations	Average: 0.28–0.48°C Max.: −0.05–0.1°C Min.: 0.38–0.8°C
55	Medium Size Cities, China	1961–2000	Standard Fixed Measuring Stations	39 years	29 urban stations	Average:0.05–0.25°C Max.: −0.15–0.0°C Min.: 0.38–0.65°C
56	Small Size Cities, China	1961–2000	Standard Fixed Measuring Stations	39 years stations	102 urban	Average: 0.05–0.25°C, Max.: 0.05–0.15°C Min.: 0.1–0.3°C
57	Six cities in the Liaoning Province of Northeast China	1980–2009	Standard Fixed Measuring Stations	29 years	24 stations	Annual: 0.57–2.15°C Monthly: −0.70–4.60°C
58	Guangzhou, China	1960–2005	Standard Fixed Measuring Stations	45 years	Not known	Average annual: 0.4–1.4°C
59	Nanjing, China	2005	Standard Fixed Measuring Stations	3 months	4 stations	Average: 0.5–3.5°C
60	Shenzhen, China	2004	Standard Fixed Measuring Stations	1 year	19 stations	Mean annual: 2.6°C
61	Xi'an, China	2005	Mobile station	Winter period	280 locations	Winter daytime: 2.5–4.0°C Winter Nighttime: 5.5–6.0°C
62	Urban areas around Yangtze River Delta, China	1961–2005	Standard Fixed Measuring Stations	44 years	99 stations	Average Annual: 0.2–0.9°C
63	Hong Kong, China	2002	Non-Standard Miniature Sensors	4 weeks	6 stations	Daytime Variation: 1.0–1.5°C
63b	Hong Kong, China	2003	Non-Standard Miniature Sensors	6 months	12–15	Daytime Variation: −1.3–3.4°C
63c	Hong Kong, China	2002	Non-Standard Miniature Sensors	3 months	6 stations	Nighttime variation: 0.4–1.3°C
64	Suzhou City, China	2007	Standard Fixed Measuring Stations	6 days	41 stations	Variation: 1.0–2.2°C
65	Colombo, Sri Lanka	1969–1999	Standard Fixed Measuring Stations	30 years	3 stations	Max. Temp. during Daytime: (−1.0)–(−1.5)°C, Variation during Night Time: 0.0–1.0°C
66	Delhi, India	March 2010	Non-Standard Miniature Sensors	One month	28 stations	Max. daily: 10.7°C Average Daily:8.3°C
66b	Delhi, India	2008	Non-Standard Miniature Sensors	3 days	30 stations	Max.: 8.3°C Min.: 2.2°C
67	Guwahati, India	2009	Non-Standard Miniature Sensors and Mobile Traverses	5 months	4 stations	Max.: 2.3°C
68	Pune, India	1997	Mobile Stations	One month	170 locations	Average: 2.0–3.1°C
69	Visakhapatnam, India	2004	Mobile Stations	Several days	Many locations	Average: 2–4°C
70	Madras, Chennai, India	1987	Mobile Stations	8 days	77 points	Max.: 2.5–4.0°C

Table 1.2 *continued*

No.	City /Country	Period of Measurements	Type of Measurements	Duration of Measurements	Number of Stations	Urban Heat Island Intensity
71	Thiruvanantha-puram, India	2010	Mobile Stations	1 day	Many locations	Average: 2.4°C
72	Bhopal, India	1980s	Mobile Stations	Not known	Many locations	Max.: 6.5°C
73	Calcutta, India	1980s	Mobile Stations	Not known	Many locations	Max.: 4°C
74	Bombay, Mumbai, India	1980s	Mobile Stations	Not known	Many locations	Max.: 9.5°C
75	Vijayawada, India	1980s	Mobile Stations	Not known	Many locations	Max.: 2.0°C
76	Koshi	2007–2008	Mobile Stations	2 days	5 Locations	Max.: 2.8°C
77	Dhaka, Bangladesh	1980s	Mobile stations	Not known	Many locations	Nighttime: 0.5–6.0°C
78	Karachi, Pakistan	2008	Standard Fixed Measuring Stations	Several days	4 stations	Average: 0.3–1.6°C
79	Hyderabad, Pakistan	2008	Standard Fixed Measuring Stations	Several days	2 stations	Average: 1°C
80	Kuala Lumpur, Malaysia	2004	Standard Fixed Measuring Stations and mobile traverses	Several days	12 stations	Daily Variation: 3.9–5.5°C
80b	Kuala Lumpur, Malaysia	1985	Standard Fixed Measuring Stations	Several days	2 stations	Max.: 4–6.0°C
81	Various cities in Klang Valley, Malaysia	1970s and 1980s	Standard Fixed Measuring Stations	Several days	Several stations	Max.: 2–5°C
82	Georgetown, Malaysia	1970s and 1980s	Standard Fixed Measuring Stations	Several days	Several stations	Max.: 4°C
83	Johor Bahru, Malaysia	1970s and 1980s	Standard Fixed Measuring Stations	Several days	Several stations	Max.: 3°C
83b	Johor Bahru, Malaysia	2008	Mobile Station	2 months	Many locations	Nighttime Max.: 4°C
84	Kota Kinabalu, Malaysia	1970s and 1980s	Standard Fixed Measuring Stations	Several days	Several stations	Max.: 3°C
85	Muar, Malaysia	July 2011	Non-Standard Miniature Sensors	31 days	3 stations	Nightime: 3.2°C Daytime: 4°C
86	Singapore	1979–1981	Standard Fixed Measuring Stations and Mobile Stations	Several months	Many locations and 4 stations	Max.: 2.5–5°C
86b	Singapore	1965	Handheld Thermometers and a Standard Measuring Station	One day	5 stations	3°C
86c	Singapore	1996–1997	Handheld Thermometers and Standard Measuring Stations	28 days	21 stations	2.5–4.8°C
86d	Singapore	2002	Mobile Station	1 day	Many locations	1.6°C
86e	Singapore	2002	Mobile Station	1 day	Many locations	4°C
86f	Singapore	2003–2004	Standard Fixed Measuring Stations	1 year	5 stations	4.3–7.1°C

Table 1.2 *continued*

No.	City /Country	Period of Measurements	Type of Measurements	Duration of Measurements	Number of Stations	Urban Heat Island Intensity
87	Ulaanbaatar, Mongolia	1980–2010	Standard Fixed Measuring Stations	31 years	2 stations	Average: 1.6°C Max.: 2.5–6.4°C
88	Jeddah, Saudi Arabia	1986	Mobile Station	One year	Many locations	0.4–2.8°C
89	Al-Hassa, Saudi Arabia	2010	Standard Fixed Measuring Stations	3 months	11 stations	Max. of Min. Temp. Differences: 9.8°C Max. of Max. Temp. Differences: 4°C Max. on Average Temp. Differences: 6.6°C
90	Mosul, Iraq	2007	Mobile Station	16 days	156 points	1–7°C
91	Teheran, Iran	1973–2003	Standard Fixed Measuring Stations	31 years	3 stations	Max. of Min. Temp. Differences: 5–7°C Max. of Max. Temp. Differences: 2–4°C
92	Beer Sheva, Israel	1965, 1979	Mobile Station	Several days	Many locations	Nighttime: 3–5°C
93	Ashdod, Israel	1970	Mobile Station	2 days	Many locations	Nighttime: 1–2°C
94	Netanya, Israel	1980	Mobile Station	2 days	Many locations	Nighttime: 1.2°C
95	Tel Aviv, Israel	1980	Mobile Station	1 day	Many locations	Nighttime: 3–5°C
95b	Tel Aviv, Israel	1996	Standard Fixed Measuring Stations and Mobile Stations	1 day	6 fixed stations	1–6°C
96	Eilat, Israel	2001–2002	Mobile Station	9 days	15 measurement points	0 3–1.6°C
97	Izmir, Turkey	1935–1990	Standard Fixed Measuring Stations	55 years	2 stations	Max.: 8°C
98	Adana, Turkey	1935–1990	Standard Fixed Measuring Stations	55 years	2 stations	Max.: 9°C
99	Bursa, Turkey	1935–1990	Standard Fixed Measuring Stations	55 years	2 stations	Max.: 9°C
100	Gaziantep, Turkey	1935–1990	Standard Fixed Measuring Stations	55 years	2 stations	Max.: 7°C
101	Ankara, Turkey	1955–1990	Standard Fixed Measuring Stations	35 years	5 stations	Average: 0.56–2.9°C

Table 1.3 Climatic characteristics that favour the development of UHI in Europe (Santamouris, 2007)

City	Impact of wind	Impact of cloud cover	Impact of cyclonic and anticyclonic conditions	The period that the UHI develops
Southern Europe				
Athens, Greece	Heat island intensity is increasing under a high pressure ridge that is characterised by weak pressure gradient and weak, variable winds or calms		High pressure ridge and close anticyclone conditions characterised by the presence of a closed anticyclone accompanied by weak winds from the southern and northern sector	Summer period
Rome, Italy	The wind action, even if not covering the effect of the urban heat island, determines a reduction of UHI differences with respect to those that could be recorded in the case of wind calm conditions			The heat island clearly appears either during the winter months or during the summer months with growing values from the winter toward the summer except for December, when a second maximum is found
Parma, Italy				The difference varied seasonally with the maximum in spring and summer
Florence, Italy	The maximum heat island intensity was found during calm days	The maximum heat island intensity was found during clear days		Summer period
Lisbon, Portugal	Weather types with northerly winds are associated with the highest air temperature in the downtown area, partly because of a shelter effect	Analysis performed only for cloudless nights		The seasonal variation of UHI frequency showed a maximum in wintertime
Aveiro, Portugal	The intensity of the island is at its maximum when there is no wind	The intensity of the island is at its maximum when the sky is totally clear	Weak UHI is associated with low pressure (cyclonic) or perturbation – atmospheric instability, strong winds and cloudiness, the occurrence of precipitation. The high-intensity islands correspond to high-pressure (anticyclonic) situations – clear sky and no wind	
Madrid, Spain	Higher UHI corresponds to low wind speeds	The intensity of the heat island is at its maximum when the sky is totally clear	Classification of UHI values on the basis of different types of weather shows that the highest values correspond to anticyclonic situations during the cold period,	Maximum at the summertime

Table 1.3 *continued*

City	Impact of wind	Impact of cloud cover	Impact of cyclonic and anticyclonic conditions	The period that the UHI develops
			and that the lowest correspond to other situations occurring during this period	
Granada, Spain	For the low wind situations, the maximum differences are higher	In the clear sky situations, the maximum differences are higher.		The maximum differences occur during winter months.
Izmir, Adana, Bursa and Gaziantep, Turkey				Urban warming is detected to be more or less equally distributed over the year, with a slight increase in the autumn months
Central Europe and UK				
Bucharest, Romania	The highest UHI is under low wind speeds	The highest UHI is under clear skies		Heat island develops better in spring and summer than in winter anticyclone situations
Szeged, Hungary	Calm or slight wind were favourable for a strong development of the heat island effect	Little or no cloud coverage were favourable for a strong development of the heat island effect	Anticyclonic weather situations were favourable for a strong development of the heat island effect.	
Debrecen, Hungary	Strong heat islands developed under anticyclonic weather conditions with weak wind speeds	Strong heat islands developed under anticyclonic weather conditions with clear skies and weak winds	Under anticyclonic conditions strong heat islands formed, but their shape was usually deformed by the prevailing winds. Strong cyclonic activity eliminated the formation of the UHI while under weak cyclonic activity regular, but weak heat islands developed	In the non-heating season, stronger heat islands develop
Wroclaw, Poland	An increase of wind speed to over 4m/s at night, and over 1m/s during the daytime, irrespective of cloudiness, totally eliminates the UHI or causes a considerable reduction in its intensity (<1 K).	The impact of cloudiness is practically unnoticeable during the daytime. At night only an increase in cloudiness to greater than 6 oktas is seen to diminish the UHI intensity		The highest average UHI intensity values are observed in the warm season, mainly in the spring (May, April)
Lodz, Poland	At night the UHI reaches its largest intensity in windless and cloudless conditions	The greatest differences occur during summer nights when skies are clear	The exceptionally intense UHI, which lasted for most of the night, is associated to the advection of the cold arctic air in which an anticyclone developed	

Table 1.3 *continued*

City	Impact of wind	Impact of cloud cover	Impact of cyclonic and anticyclonic conditions	The period that the UHI develops
Prague, Czech Republic			The increase in the heat island intensity is steeper under anticyclonic than cyclonic conditions in all seasons except for spring	The UHI increases in all seasons as well as annually; the increase is significant in all seasons except for winter when the trend is by far the smallest
London, UK				Measurements only in summertime
Northern Europe				
Moscow, Russia	Synoptic analysis of the period between May and August confirmed that low wind-speed conditions associated with anticyclones generate strong heat islands	Synoptic analysis of the period between May and August confirmed that clear-sky conditions associated with anticyclones generate strong heat islands		The highest intensities of the heat island were observed between May and August
Getemborg, Sweden	The heat island intensity is at least 2.5°C when the wind speed less than 3 m/s	The heat island intensity is at least 2.5°C when the sky is clear		Measurements performed only under anticyclonic conditions

Coastal cities experience a weaker urban heat island intensity because of the sea breezes that transfer cool air into the interior of the city (Sakakibara & Matsui, 2005). As it concerns the impact of humidity on the development of urban heat island, it is accepted that the magnitude of the phenomenon is seriously reduced during rainy days (Chow & Roth, 2006), while low relative humidity in the urban environment usually corresponds to higher UHI intensities, (Borbora & Das, 2014).

As it concerns the seasonal variation and development of the UHI, it is well accepted that the phenomenon mainly develops during the warm period of the year (Morris *et al.*, 2001), while in dry and humid climates during the dry season (Kim & Baik, 2005). During the day, UHI presents its maximum 3–5 hours after sunset (Oke, 1987), however many experiments report the measurement at various periods of the day, even during the daytime period (Skoulika *et al.* 2014).

Table 1.3 summarises the main findings regarding the climatic conditions and the period that influence the development of the urban heat island phenomenon in Europe (Santamouris, 2007).

Although the number of the experimental urban heat island studies is quite high and new data are added constantly, many inconsistencies regarding the characteristics of the studies still exist. Discrepancies between the existing information have to do with the experimental protocol used, the type and the format the results are reported in, the length of the experimental period, the number of stations used and the selection of the reference rural station.

Experimental studies on the characteristics of the canopy layer urban heat island are performed and data are collected using three specific protocols (Voogt, 2014):

1. Studies using standard meteorological stations and equipment;
2. Studies using nonstandard meteorological stations and equipment; and
3. Studies using mobile traverses.

In parallel, urban heat island studies report the magnitude of the phenomenon under different formats and calculation basis. Usually, studies based on mobile traverses report the maximum temperature difference measured during the experimental period, while studies based on standard or nonstandard meteorological stations report either the average or the average maximum temperature difference or the absolute maximum temperature difference calculated for the corresponding experimental period. The difference between the different reporting formats is significant. For example as reported by Santamouris (2015b), for all existing urban heat island studies in Asia and Australia, studies based on the average annual temperature difference report an average UHI magnitude close to 1 K, while the corresponding UHI intensity reported by studies based on the average annual maximum temperature difference and the absolute maximum temperature difference are 3.6 K and 6.2 K.

The experimental length of the urban heat island studies varies mainly as a function of the monitoring protocol used. As reported by Santamouris (2015b), about 13 per cent of the mobile traverses experiments carried out in Asia and Australia are performed for just one day while almost 49 per cent of the studies are performed for less than ten days. In parallel, the duration of all studies based on standard and nonstandard meteorological stations is much higher. About 38 per cent of the monitorings are performed for less than 30 days and 63 per cent for less than 90 days for all studies using nonstandard meteorological stations in Asia and Australia, while when standard meteorological stations are used, almost 36 per cent of the studies are extended for less than one year and 14 per cent for less than 10 years.

A complete and full knowledge of the urban heat island characteristics in a place requires a good distribution of stations in and around the city. Unfortunately, most of the studies are based on a limited number of meteorological stations that do not allow a comprehensive understanding of the phenomena in the specific place. It is characteristic that about 43 per cent of the studies performed in Asia and Australia are based on just one fixed urban meteorological station (Santamouris, 2015b).

The selection of the reference rural or suburban meteorological station is of a vital importance in urban heat island studies. Selection of an unsuitable reference station biases the urban–rural temperature differences and results in a wrong estimation of the urban heat island intensity (Morris et al., 2001). Two are the main uncertainty sources regarding the procedures followed to select the reference station:

1. In most of the large cities, important urbanization processes occurred during the last years. The possible expansion of the cities can influence the thermal balance of the traditionally considered rural zones and increase local temperatures. It is characteristic that because of the intensive urbanization,

the warming trend of the rural stations in China is close to 0.24 K per 30 years (Karl & Quayle,1988).

2. Topography around cities is not always the same, and local wind and thermal phenomena influence highly the temperature distribution around the cities. To balance the local phenomena, several stations around the city may be selected and the average value may be used as the reference (Sakakibara & Owa, 2005).

The impact of urban warming

Urban warming affects the quality of life of urban dwellers. Higher urban temperatures affect the energy consumption of buildings, increase the concentration of pollutants and deteriorate the environmental quality in the city, decrease the indoor and outdoor comfort levels, and affect human health. Important research has been carried out the last years to identify and quantify the impact of urban warming on human life in cities. The main findings are summarised below.

Impact of the urban warming on energy

High urban temperatures increase the energy needs for cooling, decrease the needs for heating and affect the peak electricity demand during the summer period. Numerous studies have been performed to investigate the topic and quantify the exact impact of urban heat island and climate change on the energy demand and consumption in cities. According to Santamouris (2015a), and considering the specific objectives of the existing analyses, five groups of relevant studies may be identified:

A. Scientific works aiming to analyse the impact of urban warming on the peak electricity demand of a city or a greater geographical zone.
B. Scientific works aiming to analyse the impact of urban warming on the total electricity consumption of a city or a country.
C. Scientific works aiming to analyse the increase or decrease of the energy consumption of typical buildings caused by the urban heat island, using meteorological data from urban and reference rural climatic stations.
D. Scientific works aiming to analyse the energy impact of the urban heat island on the total building stock in a city or a greater geographical zone.
E. Scientific works aiming to calculate the variation of the total energy consumption and the temporal energy penalty to buildings caused by the global urban warming, involving long series of past climate data.

Group A – The impact of urban warming on the peak electricity demand

There is a well-established relationship between the electricity demand during the whole year and the ambient temperature. It has the form of a U-shaped curve presenting its maximum during the warm or cold period according to the local climatic conditions. Such a curve is known as the 'response function' curve while the threshold temperature over which the electricity consumption for cooling reasons starts to increase is known as the inflection point. Usually, it is taken close to 18.3°C; however, it may change as a function of the thermal status of the electricity network. An analysis of the existing information on the relationship between ambient temperature and peak electricity demand for eleven cities and regions during the warm period is analysed by Santamouris *et al.* (2015). The main data of the analysed studies are given in Table 1.4. It is found that the increase of the peak electricity demand per degree of temperature rise varies between 0.45 to 4.6 per cent, while the average increase of the peak electricity demand per person is close to 21 (±10, 4) W per degree of temperature rise.

Table 1.4 Results of analyses on the impact of ambient temperature on the increase on the peak electricity demand (Santamouris et al., 2015)

No	City/Country	Percentage increase of the base electrical load per degree of temperature increase	Threshold inflection temperature (°C)
1	Tokyo, Japan	0.45 %	22°C
2	Thailand	4.6 % (peak demand)	—
3	Ontario, Eastern Canada	1.5 %	23°C
4	Los Angeles, CA, USA	3.3 %	18.3°C
5	Washington, DC, USA	3.6 %	—
6	Dallas/Fort Worth, TX, USA	3.1 %	13°C
7	Colorado Springs, CO, USA	1.8 %	13°C
8	Phoenix, AZ, USA	3.6 %	24°C
9	Tuscon, AZ, USA	1.8 %	21°C
10	Israel	2.9–3.1 %	—
11	Part of Carolina, USA	3.5–4 %	18°C
1	Spain	1.6 %	18°C
2	Athens, Greece	4.1 %	22°C
3	New Orleans, LA, USA	3 %	22°C
4	Hong Kong, China	4 %	18°C
5	Ohio, USA	7.5 %	16°C
6	California: San Jose, Sacramento, Pomona and Fresno	2.9 %	15°C
7	Greece	1.1 to 1.9 %	18.5°C
8	Chicago, IL, USA	Not reported	15–17°C
9	California, USA	7.7 %	17°C
10	Louisiana, USA	8.5 %.	20°C
11	Maryland, USA	8.5 %	15.6°C for residential and 11.7°C for commercial
12	Massachusetts, USA	About 6.5 % for the residential, and 3.0 % for the commercial sector	15.5°C for the residential and 12.8°C for the commercial sector
13	Bangkok, Thailand	7.49 %	—
14	Singapore	1–2.5 %	Varies during the day
15	Netherlands	0.5 %	Variable and below 18°C

Group B – The impact of urban warming on the total electricity consumption of a city or a country:

This specific group of studies aims to investigate the increase the electricity consumption per degree of ambient temperature rise. The main outcome of these specific investigations is a correlation of the electricity consumption against the corresponding ambient temperature, and an evaluation of the sensitivity of the local electricity system to the increase of the ambient temperature. Santamouris *et al.* (2015) have summarised data from relevant studies performed for fifteen cities, states and countries, around the world. The specific results are given in Table 1.5. It is found that the increase of the hourly, daily or monthly electricity consumption ranges between 0.5 per cent to 8.5 per cent per degree of temperature rise (Santamouris *et al.*, 2015). As expected, urban areas with high penetration of air condition like some states in the USA present much higher sensitivities.

Group C – Variation of the energy consumption of typical buildings caused by the urban heat island

This is the most common type of energy studies related to urban heat islands. A reference or more typical buildings are selected and their energy load is calculated under the same boundary and operational conditions using urban

Table 1.5 Increase of the electricity consumption – Impact of 1 K increase of the ambient temperature (Santamouris et al., 2015)

No (°C)	City/Country	Percentage Increase of the base electricity load per degree of temperature increase	Threshold Inflection Temperature
1	Spain	1.6 %	18°C
2	Athens, Greece	4.1 %	22°C
3	New Orleans, LA, USA	3 %	22°C
4	Hong Kong, China	4 %	18°C
5	Ohio, USA	7.5 %	16°C
6	California: San Jose, Sacramento, Pomona and Fresno	2.9 %	15°C
7	Greece	1.1 to 1.9 %	18.5°C
8	Chicago, IL, USA	—	15–17°C
9	California, USA	7.7 %	17°C
10	Louisiana, USA	8.5 %.	20°C
11	Maryland, USA	8.5 %	15.6°C for residential and 11.7°C for commercial
12	Massachusetts, USA	About 6.5 % for the residential, and 3.0 % for the commercial sector	15.5°C for the residential and 12.8°C for the commercial sector
13	Bangkok, Thailand	7.49 %	—
14	Singapore	1–2.5 %	Varies during the day
15	Netherlands	0.5 %	Variable and below 18°C

and rural climatic data. The difference between the two values is considered as the energy penalty induced by the urban heat island. Santamouris (2014), has analysed 13 studies performed in Athens, London, Munich, Rome, Volos, Boston, New York, California, Texas, Melbourne, Australia and Bahrain, comparing the cooling load of reference buildings under urban and rural temperatures. As expected, the difference of the cooling loads attributed to the urban heat island is a strong function of the characteristics of the reference buildings, the local climate and the strength of the UHI phenomenon. A comparative plot of the corresponding urban–rural cooling loads is given in Figure 1.1. As shown, the average calculated energy penalty for cooling purposes induced by the urban heat island is close to 13 per cent. The summer penalty in cooling dominated climates was found to be much higher than the corresponding decrease of the heating load.

In heating dominated urban zones, i.e. areas with an average annual temperature lower than 23°C, the decrease of the heating demand is found to be much higher than the corresponding increase of the cooling load. In parallel, it finds that the annual cooling penalty is well correlated against the logarithm of the corresponding annual cooling demand.

Group D – Energy impact of the urban heat island on the total building stock

It is evident that studies examining the cooling penalty induced by the urban heat island in a single reference building may not be representative of the whole

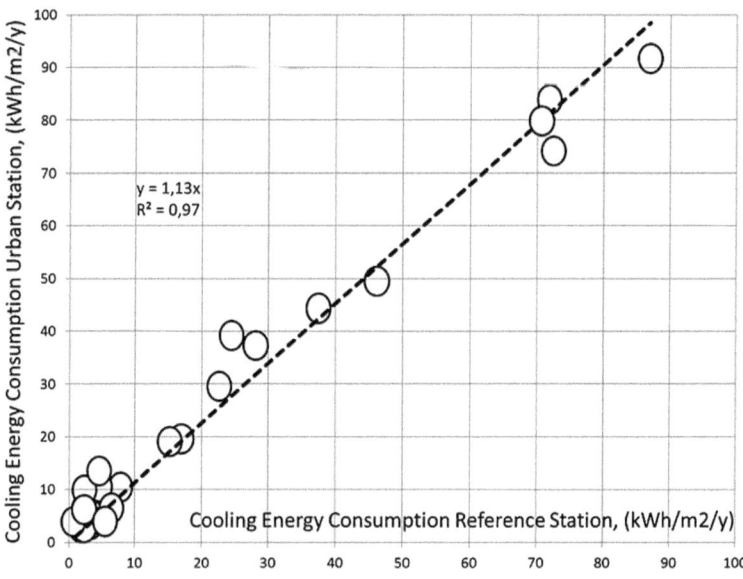

Figure 1.1 The cooling energy demand of reference buildings as calculated for the rural and the urban stations

building stock in a city. Several studies have been performed in order to evaluate the global energy impact of urban warming on the total building stock of a city or a greater region. Information and studies are available on the Municipality of Athens, Western Athens, Tokyo and Beijing. An analysis of the specific studies is performed in Santamouris (2014). It is found that the Global Energy Penalty per unit of city surface, GEPS, varies between 1.1 kWh/m² to 5.5 kWh/m². In parallel, the Global Energy Penalty per unit of city surface and per degree of the average UHI intensity, GEPSI, varies between 0.24 kWh/m²/K to 2.2 kWh/m²/K. Additionally, the Global Energy Penalty per person, GEPP, ranges between 104 kWh to 405 kWh. Thus, it may be concluded that UHI is responsible for an average energy penalty per unit of city surface around to 2.4 (±1.5) kWh/m², a Global Energy Penalty per city surface and degree of the UHI intensity close to 0.74 (±0.67) kWh/m²/K and a Global Energy Penalty per person, around for 237 (±130) kWh/p.

Group E – Variation of the total energy consumption and the temporal energy penalty to buildings

Increase of the urban ambient temperatures caused by the urban heat island and the global climate change has increased the cooling load of buildings during the last 40–50 years. Several studies have evaluated the temporal variation of the cooling load of reference buildings during this period. In particular, studies are available in Zurich, Geneva, Lugano and Davos in Switzerland; Athens, Larisa, Corfu and Heraklion in Greece; Nicosia, Paphos, Limassol, Larnaka, Famagusta and Kerynia in Cyprus; Phoenix, Washington, DC, and Puerto Rico, in the US; Hong Kong in China; and Resolute in Canada. Santamouris (2014) has analysed the results of the above studies and it is found that the cooling load of the

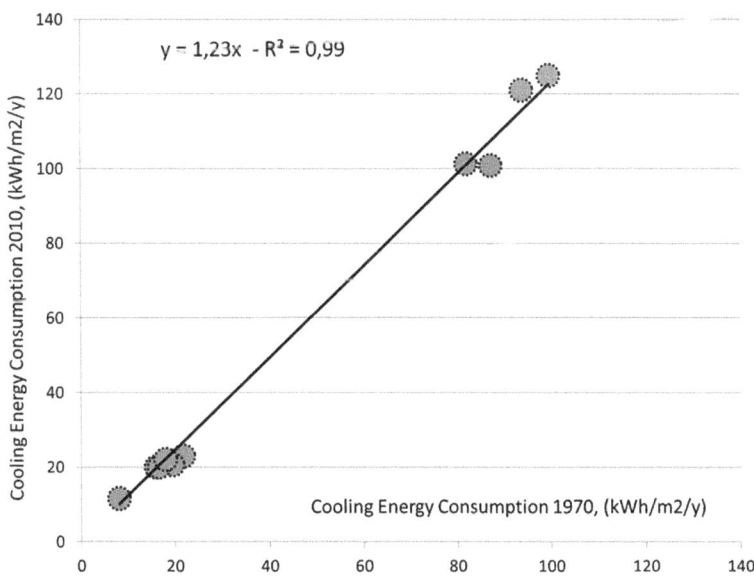

Figure 1.2 Cooling energy consumption as calculated for 1970 against the corresponding consumption for 2010

Source: Reprinted from Santamouris, M. On The Energy Impact of Urban Heat Island and Global Warming on Buildings, *Energy and Buildings*, Volume 82, October 2014, Pages 100–113, with permission from Elsevier

considered buildings has increased between 1970 and 2010 by about 23 per cent (Figure 1.2), while the total heating and cooling load during the same period has increased by 11 per cent.

Impact of urban warming on the environment, thermal comfort and health

Higher ambient temperatures in urban areas increase the ecological footprint in cities and cause a global environmental degradation. Studies performed for the city of Athens, Greece (Santamouris et al., 2007) have shown that the additional ecological footprint necessary to compensate the environmental impact of the urban heat island is equivalent to 1.5–2 times the city's political area and extends up to 110,000 hectares.

Urban warming speeds up the photochemical reactions in the atmosphere and increases the formation of tropospheric ozone (Akbari, 1992). Measurements of the tropospheric ozone in Athens during the summer period have shown that high urban temperatures have a serious impact on the concentration of this harmful pollutant (Stathopoulou et al., 2008), and days with ozone concentrations higher than the threshold value count for about 18 per cent of the summer period. Similar conclusions are also drawn from similar experiments carried out in Taiwan and Paris, France. In parallel, urban heat islands are found to slow down the penetration of the sea breeze in cities and contribute to a blocking of the pollutants (Yoshikado and Tsuchida, 1996).

The impact of urban warming on indoor and outdoor comfort conditions is very significant. Several studies have investigated indoor comfort conditions during periods of very high ambient temperatures. It is concluded that urban warming highly affects indoor comfort conditions in low income households (Sakka et al., 2012). In many cases, indoor temperatures are found to exceed the maximum allowed indoor temperatures for health reasons. In particular, measurements of indoor temperature in fifty low income houses in Athens, Greece have shown spells above 30°C measured in houses for more than 216 continuous hours (Sakka et al., 2012).

Four groups of specific studies concerning the impact of urban warming on outdoor thermal comfort are identified (Santamouris 2015a). The first cluster of investigations examines the impact of urban heat island on the spatial distribution of the outdoor thermal comfort levels. All existing studies conclude that urban zones suffering from UHI present much lower thermal comfort levels, while in some cases, dangerous discomfort conditions are monitored (Giannopoulou et al., 2013). The second cluster of studies investigates the evolution of the outdoor thermal comfort in a specific place because of the continuous interannual temperature increase. It is characteristic of a study on the evolution of the outdoor thermal comfort conditions in Athens, Greece, during the period 1954–2012, which has shown that after 1980 the intensity and frequency of thermal discomfort conditions in the city considerably increased and the frequency of high discomfort days during this period has doubled (Bartzokas et al., 2013). The third cluster of articles involves investigations of the prevailing outdoor thermal comfort conditions during extreme weather conditions. A study performed also in Athens, Greece has evaluated thermal comfort conditions during the period of a heat wave (Papanastasiou et al., 2015). It is found that during this period half of the population in the city was under severe heat stress and for a substantial part of the heat wave, the population was under medical emergency conditions. Finally, the fourth cluster of studies investigates the future outdoor thermal comfort conditions in various zones of the planet because of the global warming. All studies concluded that thermal discomfort conditions may increase in both duration and frequency in the immediate future (Santamouris, 2015a).

It is widely accepted that high urban temperatures have a serious impact on human health, increasing hospital admissions during the warm periods and heat-related human mortality (World Health Organisation, 2007). Important medical research has found that high ambient temperatures decrease the viscosity of blood and increase the risk of thrombosis, may cause important cardiovascular and respiratory problems, cerebrovascular disorders, and cause problems of thermoregulation and impaired kidney function (Flynn et al., 2005). In parallel, it is reported that higher urban temperatures above 27°C increase problems related to mental and behavioural disorders (Hansen et al., 2008).

Statistical data from many places around the world have shown that during periods of extreme heat, there is an important increase in the hospital admissions related to heat exhaustion and stroke (Pirard et al. 2005). As reported by Loughnan et al. (2010), during the period of heat waves in Melbourne, Australia, hospital admissions because of acute myocardial infractions increased between 10.8 to 37.7 per cent as a function of the specific climatic conditions.

In parallel, high temperatures in the urban environment increase mortality rates and cause more fatalities. During the 2003 heat wave in France, while the heat-related excess mortality in small cities was close to 40 per cent, the corresponding figure in Paris was 141 per cent (Dousset et al., 2010). The relationship between the excess heat related mortality and ambient temperature follows a U-shaped curve where heat related mortality increases rapidly over a threshold temperature varying as a function of the local climate, and the physiological characteristics of the population (Wilkinson et al., 2001). Recent studies carried out in various European cities have shown that the threshold temperature in the Mediterranean cities was 29.4°C, while it was much lower, 23.3°C, in northern and continental cities (Baccini et al., 2008). The increase in rate of mortality above the threshold temperatures values and per degree of increase of the ambient temperature was 3.12 per cent in the Mediterranean and 1.84 per cent in Northern Europe.

References

Akbari, H. (1992). *Cooling our communities: a guidebook on tree planting and light colored surfaces*. Berkeley, CA: Lawrence Berkeley National Laboratory.

Baccini, M., Biggeri, A., Accetta, G., Kosatsky, T., Katsouyanni, K., Analitis, A., Anderson, H.R., Bisanti, L., D'Ippoliti, D., Danova, J., et al. (2008). Heat effects on mortality in 15 European cities. *Epidemiology 19*, 711–719.

Bartzokas, A., Lolis, C. J., Kassomenos, P. A., and McGregor, G. R. (2013). Climatic characteristics of summer human thermal discomfort in Athens and its connection to atmospheric circulation. *Natural Hazards and Earth System Science 13*, 3271–3279.

Borbora, J. and Das, A.K. (2014). Summertime urban heat island study for Guwahati City, India. *Sustainable Cities and Society 11*, 61–66.

Brázdil, R. and Budíková, M. (1999). An urban bias in air temperature fluctuations at the Klementinum, Prague, the Czech Republic. *Atmospheric Environment 33*(24–25), 4211–4217.

Chow, W.T.L., and Roth, M. (2006). Temporal dynamics of the urban heat island of Singapore. *International Journal of Climatology 26*, 2243–2260.

Dousset, B., Gourmelon, F., Laaidi, K., Zeghnoun, A., Giraudet, E., Bretin, P. ,Mauri, E., and Vandentorren, S. (2010). Satellite monitoring of summer heat waves in the Paris metropolitan area. *International Journal of Climatology 31*(2), 313–323.

Flynn, A., McGreevy, C., Mulkerrin, E.C. (2005). Why do older patients die in a heatwave? [Commentary] *QJM 98*, 227–229.

Giannopoulou, K., Livada, I., Santamouris, M., Saliari, M., Assimakopoulos, M., & Caouris, Y., (2013). The influence of air temperature and humidity on human thermal comfort over the greater Athens area. *Sustainable Cities and Society 10*, 184–194.

Hansen, A., Bi, P., Nitschke, M., Ryan, P., Pisaniello, D., Tucker, G. (2008). The effect of heat waves on mental health in a temperate Australian city. *Environmental Health Perspectives 116*, 1369–1375.

IPCC. (2014). Summary for policymakers. In: *Climate Change 2014: Impacts, Adaptation, and Vulnerability*. Contribution of Working Group II to the Fifth Assessment Report of the Intergovernmental Panel on Climate Change. Intergovernmental Panel on Climate Change. Cambridge University Press, Cambridge, UK.

Karl, T.R.. and Quayle, R.G. (1988). Climatic change in fact and theory: are we collecting the facts? *Climatic Change 13*, 5–17.

Kim, Y., and Baik, J.-J. (2004). Daily maximum urban heat island intensity in large cities of Korea. *Theoretical and Applied Climatology 79*, 151–164.

Kim, Y., and Baik, J.-J. (2005). Spatial and temporal structure of the urban heat island in Seoul. *Journal of Applied Meteorology 44*, 591–605.

Landsberg, H.E. (1981). *The urban climate*. New York: Academic Press.

Li, D., and Bou-Zeid, E. (2013). Synergistic interactions between urban heat islands and heat waves: the impact in cities is larger than the sum of its parts. *Journal of Applied Meteorology and Climatology 52*, 2051–2064.

Loughnan, M E., Neville, N., Tapper, N.J. (2010). The effect of summer temperature, age and socioeconomic circumstance on acute myocardial infarction admissions in Melbourne, Australia. *International Journal of Health Geographics 9*(41), 134–145.

McCarthy, M.P., Best, M.J., and Betts, R. A. (2010). Climate change in cities due to global warming and urban effects. *Geophysical Research Letters 37*, L09705, doi:10.1029/ 2010GL042845.

Mihalakakou, G., Flocas, H.A., Santamouris, M., and Helmis, C.G. (2002). Application of neural networks to the simulation of the heat island over Athens, Greece, using synoptic types as a predictor. *Journal of Applied Meteorology 41*(5), 519–527.

Morris, C.J.G., Simmonds, I., and Plummer, N. (2001). Quantification of the influences of wind and cloud on the nocturnal urban heat island of a large city. *Journal of Applied Meteorology 40*, 169–182.

Oke, T.R. (1987). *Boundary Layer Climates*. London and New York: Methuen.

Oke, T.R. (1997). Urban climates and global change. In: A. Perry and R.Thompson (eds). *Applied climatology: principles and practice* (pp. 273–287). London: Routledge.

Oke, T.R., Johnson, G.T., Steyn, D.G., and Watson, I.D. (1991). Simulation of surface urban heat islands under 'ideal' conditions at night – part 2 : diagnosis and causation. *Boundary Layer Meteorology 56*, 339–358.

Papanastasiou, D.K., Melas, D., and Kambezidis, H.D. (2015). Air quality and thermal comfort levels under extreme hot weather. *Atmospheric Research 152*, 4–13.

Park, H.C. (1986). Features of the heat island in Seoul and its surrounding cities. *Atmospheric Environment 20*, 1859–1866.

Pirard, P., Vandentorren, S., Pascal, M., Laaidi, K., Le Tertre, A., Cassadou, S., and Ledrans, M. (2005). Summary of the mortality impact assessment of the 2003 heat wave in France. *Eurosurveillance 10*, 153–156.

PWC. (2014). *Rapid urbanisation*. Available at: http://www.pwc.com/gx/en/issues/mega trends/rapid-urbanisation-ian-powell.jhtml.

Sakakibara, Y., and Matsui, E. (2005). Relation between heat island intensity and city size indices/urban canopy characteristics in settlements of Nagano Basin, Japan. *Geographical Review of Japan 78*(12), 812–824.

Sakakibara, Y., and Owa, K. (2005). Urban–rural temperature differences in coastal cities: influence of rural sites. *International Journal of Climatology 25*, 811–820.

Sakka, A., Santamouris, M., Livada, I., Nicol, F., and Wilson, M. (2012, June). On the thermal performance of low income housing during heat waves. *Energy and Buildings 49*, 69–77.

Santamouris, M. (2007). Heat island research in Europe – the state of the art. *Advances in Building Energy Research 1*, 123–150.

Santamouris, M. (2014, October). On the energy impact of urban heat island and global warming on buildings. *Energy and Buildings 82*, 100–113.

Santamouris, M. (2015a). Regulating the damaged thermostat of the cities – status, impacts and mitigation challenges. *Energy and Buildings 91*, 43–56.

Santamouris, M. (2015b). Analyzing the heat island magnitude and characteristics in one hundred Asian and Australian cities and regions. *Science of the Total Environment 512–513*, 582–598.

Santamouris, M., Cartalis, C., Synnefa, A., and Kolokotsa, D. (2015). On the impact of urban heat island and global warming on the power demand and electricity consumption of buildings – a review. *Energy and Buildings*. doi: 10.1016/j.enbuild.2014.09.052

Santamouris M., Paraponiaris, K., and Mihalakakou, G. (2007). Estimating the ecological footprint of the heat island effect over Athens, Greece. *Climate Change 80*, 265–276.

Skoulika, F., Santamouris, M., Boemi, N., and Kolokotsa, D. (2014). On the thermal characteristics and the mitigation potential of a medium size urban park in Athens, Greece. *Landscape and Urban Planning 123*, 73–86.

Stathopoulou, E., Mihalakakou, G., Santamouris, M. and Bagiorgas, H.S. (2008, June). Impact of temperature on tropospheric ozone concentration levels in urban environments. *Journal of Earth System Science 117*(3), 227–236.

United Nations. (2014). *World urbanization prospects*. 2014 Revision.

Voogt, J. (2014). How researchers measure urban heat islands. Available at: http://www.epa.gov/heatislands/resources/pdf/EPA_How_to_measure_a_UHI.pdf.

Wilkinson, P., Armstrong, B., Fletcher, T., Landon, M., Mckee, M., Pattenden, S., and Stevenson, S. (2001). Cold comfort: the social and environmental determinants of excess winter deaths in england, 1986–96. Bristol: The Policy Press.

World Health Organisation. (2007). *Large analysis and review of European housing and health status (LARES)*. Copenhagen, Denmark: WHO Regional Office for Europe, DK-2100

Yoshikado, H., and Tsuchida, M. (1996). High levels of winter air pollution under the influence of the urban heat island along the shore of Tokyo Bay. *Journal of Applied Meteorology and Climatology 35*, 1804–1813.

2

UNDERSTANDING AND REDUCING THE ANTHROPOGENIC HEAT EMISSION

Nektarios Chrysoulakis[1] and C.S.B. Grimmond[2]

[1]Foundation for Research and Technology – Hellas – Heraklion, Greece
[2] University of Reading – Reading, UK

Introduction

Cities and the behaviour of their residents change through time. The spatial organization of urban areas, whether the footprint or height of buildings; the width, length, and layout of streets; where people live and where they work; the walkability and scale of neighbourhoods; the relative proximity of urban amenities; the type and diversity of urban green space all affect urban energy use. The global urban population is expected to increase to up to 7 billion by 2050 (United Nations 2010). Thus there are enormous opportunities to shape the built environment and for urban planning to play an important role in climate change mitigation (implementation of mitigation technologies, such as green roofs and cool materials) and adaptation (measures such as emergency management plans) at the city level (Santamouris 2014). Mitigating climate change can be achieved better by regulating land-use change than by CO_2 emissions reductions alone (Stone 2009). The regionalized impacts of the anthropogenic heat emissions on climate have the effect of more directly localizing the benefits of land-based mitigation (Stone *et al.* 2012).

One important challenge facing the urbanization and global environmental change community is to understand the relations between urban form, energy use and carbon emissions at different spatial and temporal scales. Yet, these types of analyses are needed by urban planners, who recognize that zoning a city at the local scale affects mobility and transportation choices, energy consumption and climate. As indicated by Seto and Christensen (2013), what is missing from most existing climate action plans is a clear link between urban land-use patterns beyond the individual parcel or building scale and energy use.

Energy enters, passes and leaves a city in numerous ways and physical states and forms; for example, fossil fuels, electricity, radiation and heat. Construction materials, food, water and waste also store energy. What type of energy is considered depends on perspectives and applications. For example, urban planners, economists and statisticians, tend to focus on fluxes of energy that are 'usable' for day-to-day activities and the optimization of this, with concern for questions such as how to influence energy consumption by administrative means, such as guidelines or regulation for insulation of new houses, traffic reduction, etc. Meteorologists, in contrast, are more interested in understanding how energy in forms such as radiation and heat influences the urban climate, and how it is transported and stored in the urban fabric and atmosphere. Energy flow charts by urban planners usually omit radiation and anthropogenic heat as a heating source (Chrysoulakis *et al.* 2009). Urban meteorologists, in contrast, consider them in the Urban Energy Budget (UEB), taking account of the anthropogenic heat resulting from vehicular emissions, space heating and cooling of buildings, industrial processing, and the metabolic heat release by people.

The focus of this chapter is the anthropogenic heat flux, its quantification and significance and opportunities to moderate its effect on the urban climate.

Quantifying the anthropogenic heat flux

Given the three-dimensional nature of the urban ecosystem, the UEB approach considers a control volume, i.e. all terms in the balance equation represent fluxes into, or out of, or the storage change of this volume (Figure 2.1). The surface energy balance, with units of energy (J, joules) per unit time (s, seconds) per surface area (m²), is defined for the top of the volume as:

$$Q* + Q_F = Q_H + Q_E + \Delta Q_S + \Delta Q_A + S \ (J \ s^{-1} \ m^2 = W \ m^{-2}) \qquad (1)$$

where Q^* is the net all-wave radiation, Q_F is the anthropogenic heat flux, Q_H is the turbulent sensible heat flux, Q_E is the turbulent latent heat flux, ΔQ_S is the net change in heat storage within the volume (the volume is defined such that the flux into the ground is incorporated in ΔQ_S), ΔQ_A is the net advected flux ($\Delta Q_A = Q_{in} - Q_{out}$) and S represents all the other sources and sinks (e.g. heat removed by runoff, photosynthetic heat, etc.); the heat-to-wastewater flux (Iamarino *et al.* 2012) is also included in S. This equation is applicable for small time steps (1 hour), or if the UEB is integrated over the course of a 24-hour period, the net storage term may be assumed small and ignored.

Significant advances have been made in the simulation of the physics of urban climate processes associated with enhanced computational capacity, improved

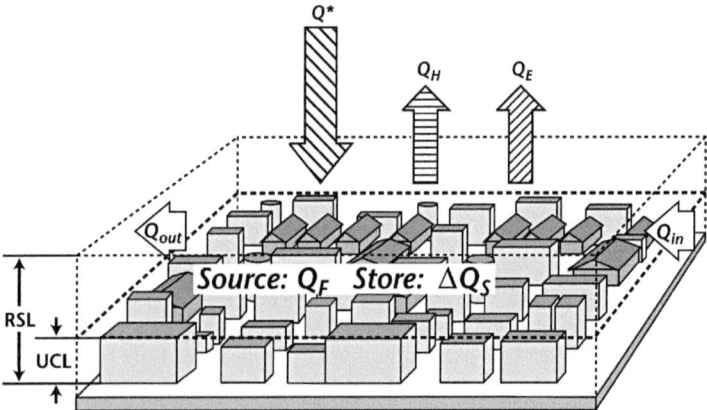

Figure 2.1 Conceptual illustration of the fluxes in the energy balance of an urban building–soil–air volume
Source: Roberts et al. 2006. UCL is the Urban Canopy Layer and RSL is the Roughness Sub-Layer.

Earth Observation (EO) sensors and model developments (Arnfield 2003, Weng 2009, Grimmond *et al.* 2010, 2011). Q_F is difficult to determine because of its strongly varying pattern and because it is very difficult to measure directly. Q_F consists of both sensible and latent heat components. Their relative importance varies not only with urban land use but also by time of year. In areas where evaporative towers are used in large cooling systems, the latent heat portion can be more than 20 per cent of total Q_F in summer; for example central Tokyo (Moriwaki *et al.* 2008) and Osaka Prefecture (Narumi *et al.* 2009).

There are three general approaches to estimate Q_F (Sailor 2011):

- *Inventories*: Commonly socio-economic data are used with energy-use data to the three main contributors: mobile (e.g. traffic), stationary sources (buildings) and metabolic (human) release. The fraction of traffic as a function of type and amount of gasoline, the number of vehicles, their fuel efficiency and the distance they travel commonly are used. For stationary sources, consumer-scale long-term data can be divided into short-term fluctuations using large-scale hourly variations of electricity/gas consumption (Grimmond 1992, Sailor and Lu 2004, Moriwaki *et al.* 2008, Hamilton *et al.* 2009, Smith *et al.* 2009). Recently, Iamarino *et al.* (2012) assessed the spatial variability of Q_F from buildings using high-resolution resident and workplace population data. They concluded that buildings are the major source of anthropogenic heat emissions, accounting for about 80 per cent of the nearly 150 TWh of waste energy annually emitted across the Greater London area. As discussed by Sailor (2011) such inventory approaches are limited by resolution. Typically, energy consumption data have good temporal resolution at coarse spatial scales, but poor temporal resolution at more detailed spatial resolution. In addition, assumptions have to be made about energy consumption and its relation to release and the location of the release (e.g. all to the external environment).
- *Residual of the UEB*: A second approach to estimate Q_F is to calculate the flux as the residual of the UEB (equation 1). This is a more physically-based method but all errors of the other terms are cumulated the residual flux. This method also assumes energy balance closure for the time interval of calculation. Measurements of the other UEB terms may have systematic biases, for example, if the turbulent heat fluxes are underestimated by the measurement technique (e.g. Eddy Covariance) or insufficient sampling (e.g. to obtain measurements to determine the storage or radiative fluxes). To simplify the problem, long time periods are used, so the storage term can be assumed to be negligible (Pigeon *et al.* 2007). Furthermore, there is the incorrect assumption that all the UEB fluxes are commonly measured at the same spatial scale. The high level of uncertainty associated with residual Q_F means both the temporal and spatial resolution are coarse (of the order 10^2 – 10^3 m) and diurnal profiles cannot be resolved. Thermal remote sensing may have the potential to improve this approach (see, for example, Kato and Yamaguchi 2005, 2007, Rigo and Parlow 2007, Kato *et al.* 2008, Xu *et al.* 2008, Chrysoulakis *et al.* 2015).
- *Building energy models:* As the release of anthropogenic heat is typically largest from the building sector, it has been the subject of many focused studies. These bottom-up studies generally involve explicit modelling of energy consumption within buildings and careful evaluation of heat emissions. Typically, prototype buildings are modelled using detailed building energy simulations. These results can then be integrated with an urban canopy meteorological model, which for completeness requires the other

two components of Q_F to be accounted for (transport and metabolism). The building energy sub-models explicitly account for building occupants, radiative transfer through windows, type of air-conditioning heat exchanger (air-cooled vs evaporatively cooled) and performance of air-conditioning systems, in addition to the building shape, materials, etc. The use of such a detailed building energy sub-model results in more realistic estimate of anthropogenic emissions (Kikegawa *et al.* 2003, Sailor 2011, Bueno *et al.* 2012). For the most part, these building-sector prototype approaches have focused on generating representative profiles of anthropogenic emissions of heat and moisture rather than broader estimates. Only recently have the models been dynamically linked with urban atmosphere models (Sailor 2011, Bueno *et al.* 2012). These can provide data at high spatial and temporal resolution but the contribution of transport is ignored.

A better understanding of local-scale interactions between typical urban units and the atmosphere is needed for both urban planning and landscaping and climate mitigation strategies. In particular, the key characteristics governing energy exchange between the surface and atmosphere have to be identified if adequate action is to be taken to improve thermal comfort, reduce energy use, or reduce the impacts of extreme heat/cold events. Such knowledge is also critical for a better characterization of boundary layer processes and therefore for applications in the fields of air quality and pollutant dispersion. In many cases the release of this heat is also associated directly (e.g. vehicles) or indirectly (e.g. buildings) with pollutant release, whether the greenhouse gas of carbon dioxide or small particulates (e.g. PM10). Thus changes in energy use and heat release may also change the pollutants that are released and/or the location of their release.

The importance of the anthropogenic heat flux

Nearly all energy used for human purposes is dissipated as heat within the Earth's surface-atmosphere system. Thermal energy released from non-renewable sources is therefore a climate forcing term. Flanner (2009) estimates that

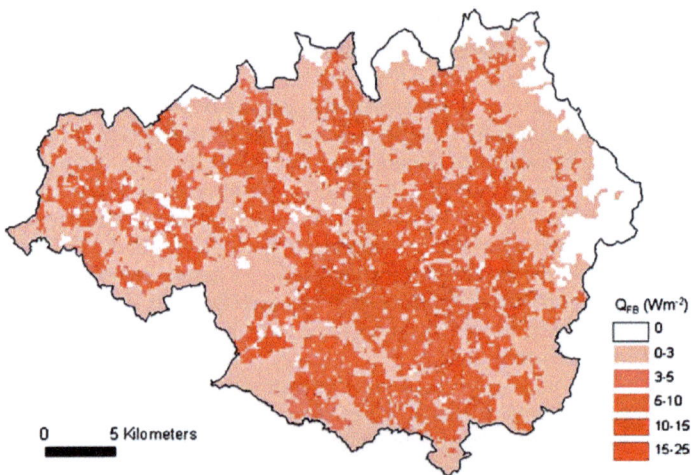

Figure 2.2 Daily averaged Q_F (W m^{-2}) estimates for Greater Manchester at 200 m x 200 m
            grid, using urban morphology units
Source: Smith et al. 2009.

averaged globally this forcing is only +0.028 W m^{-2}, but over the continental United States and Western Europe, it is +0.39 and +0.68 W m^{-2}, respectively. The anthropogenic heat flux emission is, though, strongly affected by climate and weather, because this determines whether heating or cooling systems are used. In the USA, between 60 and 70 per cent of energy consumption in buildings is used for heating, air-conditioning and water heating (Heiple and Sailor 2008). While predicting climate change and its impacts at a global scale has uncertainties, local effects of urbanization on the climate have long been documented. Recent work has documented that most large U.S. cities are warming at double the rate of the planet as a whole, a trend attributed to rapid growth in the UHI phenomenon (Stone *et al.* 2012). Urban warming has important implications for human comfort, health and well-being. Many examples exist of the vulnerability of urban populations, most often young children, the elderly and the poor, associated with heat waves; for example in Chicago, USA (1995), India (1998), France, Spain and the UK (2003) and Russia (2010). Future climate scenarios, which predict an increase in summertime maximum temperatures and also in the frequency and magnitude of extreme conditions, suggest greater risks for vulnerable people in the future. Warmer conditions in cities will also increase demand for air conditioning. More air conditioners generate more heat and have significant effects on the local-scale external climate, with implications for human comfort and the demand for cooling. At a larger scale, greater use of air conditioning results in more greenhouse gases through increased electricity generation. Concurrent with rising heat wave frequency also is a growing incidence of heat-related health effects among urban populations and more frequent failures of critical infrastructure, such as electrical power generation and transmission systems. Confronted with these trends, many urban governments have revised or developed new heat wave emergency response plans, as part of broader climate change mitigation and adaptation strategies. This has broader implications in terms of the management of energy resources to reduce Q_F.

Q_F varies widely through the year, through the day, between countries and urban areas. Although typical values in cities range from 5–100 W m^{-2} (Christen and Vogt 2004, Allen *et al.* 2011), very high Q_F values under extreme localized conditions have been reported. In a modern city core with high-rise buildings and intense traffic, Q_F can exceed 100 W m^{-2}; for example Q_F values up to 550 W m^{-2} for small areas in central London (Iamarino *et al.* 2012). Furthermore, changes in energy consumption due to changes in climate are predicted to cause a 13 per cent (11 per cent) increase in Q_F on summer (winter) weekdays in Europe (Lindberg *et al.* 2013). The largest impact results from changes in temperature conditions, which influences building energy use. The spatial resolution used to model Q_F is critical and highlights implications for scientific understanding and decision-making. High resolution databases (100 m x 100 m) typically are only developed for individual cities or small areas within them (Lindberg *et al.* 2013). At the other extreme are the global (0.5° x 0.5°) datasets; for example Flanner (2009).

In cities, peak values are associated with large buildings and busy roads. Smith *et al.* (2009) estimated Q_F spatiotemporal patterns for the city of Manchester at 200 m x 200 resolution, using a top-down approach based on urban morphology units. Their daily averaged Q_F estimates (Figure 2.2) highlighted the potential for exposure to localized heat sources which may lead to additional thermal discomfort and overheating during the summer months; and conversely the potential for lowered energy consumption due to a decrease in heating demand in urban centres during winter. Lindberg *et al.* (2013) employed the Large-scale

Figure 2.3
Daily averaged Q_F (W m^{-2}),
based on LUCY (Large-
scale Urban Consumption
of energy) model
simulations for 14th
February 2005. Population
distribution is an important
factor in the spatial
distribution of Q_F. This is
exemplified by comparing
e.g., London and Paris, two
similar cities based on size
and population, but Paris is
predicted to have a much
higher peak in Q_F, because
of higher population
density at its city centre.
Source: Lindberg et al. 2013.

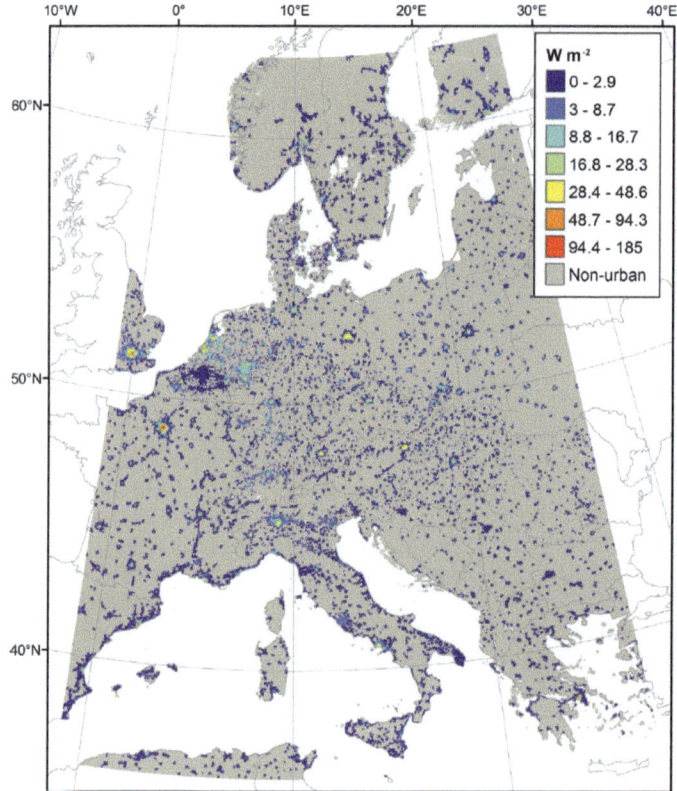

Urban Consumption of energY (LUCY) model to estimate the spatial distribution
of Q_F across Europe for the period 1995–2015, considering changes in
temperature, population and energy use. They found that while on average Q_F
was small (of the order 1.9–4.6 W m^{-2} across all the urban areas of Europe),
significant spatial variability existed (maximum 185 W m^{-2}). Therefore, while still
small at the individual sites, aggregated over the entire urban area of Europe Q_F
represents a large amount of additional energy that needs to be accounted for
in considerations of urban energy exchanges, as shown in Figure 2.3 (Lindberg
et al. 2013).

In terms of its temporal variability, as a general rule, the diurnal profile of total
anthropogenic heating has local peaks in the early morning and late afternoon,
corresponding to peaks in transportation and building-energy use. However, in
commercial and industrial areas these peaks maybe less obvious. Grimmond
et al. (2010) argue that Q_F may become especially significant in the UEB at key
times of the day and night and specifically at transition times. As illustrated in
Figure 2.4 (Sailor 2011), the diurnal profile also depends upon the day of the
week, with total emissions on weekends and holidays significantly lower than
emissions on workdays. The relative magnitude of the morning and evening
peaks depends upon the underlying climate and the season. In summer, the
afternoon peak becomes more pronounced as a result of air-conditioning loads
that tend to peak between 15:00 and 17:00 local time. In the winter, the morning
peak may become more pronounced as a result of heating demand in the building

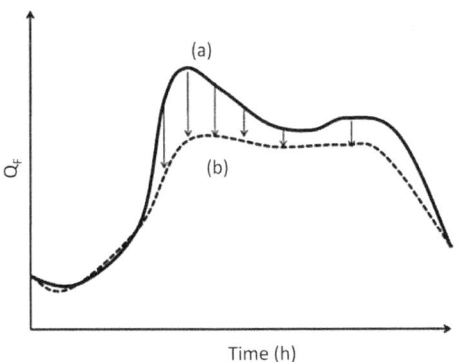

Figure 2.4
Typical shapes of diurnal profiles of anthropogenic heating for (a) work days, and (b) non-work days, illustrating local peaks in the morning and early evening hours
Source: Adapted from Sailor 2011.

sector. Therefore, producing local-scale, spatially and temporally resolved Q_F profiles would be of value not only for inclusion to, and the evaluation of, UEB models, but also for use by urban planners and designers to understand and better manage the future impacts of a changing climate on patterns of risk within urban populations.

The sensible anthropogenic heat emissions can lag the timing of the energy consumption and can differ substantially in magnitude. The bulk of the discharge from transportation is directly into the atmosphere and without significant time lag. As a consequence, profiles of energy generation from road fuels and atmospheric heat emissions are coincident, and the same applies to metabolic emissions from people outdoors. This is not true for the energy consumed in buildings and the metabolic emissions from people within because of the heat-transfer resistance between buildings and atmosphere and the thermal inertia of buildings. Moreover, the total energy exchanged by the buildings with the atmosphere can be larger than just the anthropogenic sources in buildings, because it also includes the short-wave and long-wave radiation absorbed by buildings less the heat dissipated into the ground. Because of this, the total energy exchanged by the buildings with the atmosphere is not equivalent to the energy consumption in buildings (Iamarino et al. 2012).

Anthropogenic moisture emissions also take two forms. Heat removed from buildings can be exhausted through evaporative cooling equipment. Such heat removal is often referred to as anthropogenic latent heating, with the net effect on the urban atmosphere also being a source of moisture. The second mechanism of anthropogenic moisture emissions is the chemical reaction that occurs in the combustion of hydrocarbon fuels either in vehicle engines or combustion furnaces. This process creates anthropogenic water vapour as a result of a chemical reaction rather than phase change and is thus distinct from latent heat emissions. Usually, the term anthropogenic heat is used to refer to sensible emission of heat associated with energy consumption. The latent and chemical generation of anthropogenic moisture resulting from energy consumption in the urban environment is referred to collectively as anthropogenic moisture emissions. Sailor et al. (2007) estimated that, due to significant environmental loads, heat emission from buildings can be 50 to 100 per cent greater than the energy consumption of the building. They also found that an important percentage of the heat emitted from buildings in the commercial core of a city such as Houston (with large buildings with evaporative cooling towers) is emitted as latent heat.

Anthropogenic heat flux and local climate zone

Knowledge of Q_F at the local scale is important in Local Climate Zones (LCZ) identification. Stewart and Oke (2012) formally defined LCZs as regions of uniform surface cover, structure, material, and human activity that span hundreds of metres to several kilometres in horizontal scale. LCZs simplify the broad variety of urban forms into 17 categories that are characterized by building morphology, roads, plants, soils, rock, and water. Descriptive sheets of the LCZ classes include images to enable the general urban forms to be recognized. Obviously not all urban areas fit perfectly into a LCZ and there are areas of transition. LCZ classes are local scale units, a minimum areal extent of the order of 400–1000 m, with dimensions based on the length scales that are needed for the atmosphere to develop an internal boundary layer that is adjusted in response to surface features of the LCZ. Close to the upwind edge of the LCZ, the depth of this layer will be shallow but will develop with distance into the LCZ. This scale is common to urban planning scales. Thus the LCZ system has the potential to support well-established planning projects such as climatope maps (Scherer *et al.* 1999) and urban climatic maps (Ren *et al.* 2011). To help quantify the thermal and morphological layers of an urban climatic map, standardized metadata for urban structure, cover, fabric and metabolism can be extracted from the LCZ. Mitraka *et al.* (2015) have shown that LCZ can be determined using multiple EO data sources by defining five properties: the sky view factor; the building density; the impervious and pervious surface fraction; the mean building/tree height; and the surface albedo. LCZs can help to identify the key processes governing the partitioning of energy at the urban surface and they also provide guidance as to what sort of Q_F values should be expected for a particular area where surface characteristics are known.

Methods and technologies to reduce anthropogenic heat emission

Recent research has permitted the development of technological counter-measures to the impact of urban warming. Mitigation techniques aim to balance the thermal budget of cities by increasing thermal losses and decreasing the corresponding gains. As a variety of factors (e.g. surface cover, anthropogenic heat release and urban characteristics including morphology, geographic features and climatic conditions) interact with one another to generate the UHI effect, a range of measures are needed to be effective. Among the more important of the UHI mitigation techniques are those targeting the increase of the albedo of the urban environment, to expand the green spaces in cities and to use the natural heat sinks in order to dissipate the excess heat strength (Santamouris 2014). Strategies for mitigation of UHI effects should therefore include measures related to: a) urban morphology and structure; b) urban materials; and c) reduction of Q_F.

From the perspective of urban planning and design, urban planners should focus on the determination of appropriate parameters and practices, which can be modified based on technological interventions capable of reducing Q_F. Therefore parameters (relative to buildings and urban design) and practices (relative to traffic) can be determined in each city, and technological solutions may be proposed on their modification towards reducing the anthropogenic heat. For example, parameters relative to buildings include ventilation, shading, use of appropriate covering materials and relative position with regards to heat sources. The scale at which interventions for the reduction of anthropogenic heat will be

Table 2.1 Categories of Q_F reduction measures (adapted from Yamaoto 2006)

Measure	Action
Improvement in the efficiency of energy-using products	Office automation equipment and electric consumer applications
Improvement in the efficiency of air-conditioning systems	Refrigerators and heat-source equipment
Optimal operation of air-conditioning systems	• Proper placement of outdoor units • Use of cooling towers • Voluntary restraints on night-time operations
Improvement in the heat insulation and thermo-shield of buildings	• High-performance heat insulation materials (interior heat insulation materials) • High-performance heat insulation and thermo-shield materials (exterior heat insulation materials)
Greening of buildings and adoption of water-retentive materials	Greening of buildings and adoption of water-retentive materials (exterior heat insulation materials)
Improvement in the reflectivity of walls and roofing materials	Light coloured walls and highly reflective roofing materials
Introduction of traffic-control measures	• Traffic demand management and introduction of low emission vehicles • Promotion of alternatives such as bicycles
Introduction of district heating and cooling	Central control of exhaust heat from buildings (at the regional level)
Use of untapped energy	• Use of sea, river and ground water • Use of exhaust heat from urban facilities (use of exhaust heat from industrial plants, subways, buildings, power plants, substations etc.) • Recovery of energy from waste materials (waste power generation and heat supply)
Use of natural energy	• Photovoltaic generation • Use of solar heat

implemented, should be also defined. At local scale (neighbourhood), technologies focus on urban micro-climate and urban design. At building scale, they
focus on optimization of radiation and air fluxes. At micro-scale, they focus on
thermal comfort and on impacts on human health. The selection of appropriate
technologies must be based on international and local practices and resources.
The potential of each technology for an area, for different scales, should be
evaluated. For example, at micro- or building-scale, the increase of the shading
by trees and reduction of surface temperature through greening rooftops and
walls; and at local scale, the increase of urban density with building orientations,
through the appropriate planning and design. The Q_F reduction measures that
are considered most effective, as summarised by Yamamoto (2006), are shown
in Table 2.1.

A challenging aspect of assessing the impact of Q_F on thermal comfort is the
estimation of the actual location of the emissions from buildings. Some of the
anthropogenic emissions occur as conduction through the building envelope. A
larger fraction occurs as a result of air exchanges through the facade and through
natural operation of windows and doors. The largest fraction of anthropogenic
emissions from buildings comes in the form of heat and moisture removed
by mechanical heating, cooling, and ventilation systems. It is thus important to
know where these systems are located. As discussed by Sailor (2011), in buildings
that are fewer than 10 stories tall, the heating, ventilation and air-conditioning
equipment may be located adjacent to the building on the ground level, on a
mezzanine-level roof, or on the rooftop of the top floor. In mid- and high-rise

commercial buildings, however, the emissions may be more uniformly distributed vertically on utility floors, or consolidated at the rooftop level.

Priority for implementing UHI mitigation measures has commonly been given to measures that are readily available, but there is a growing need to adopt long-term, large-scale measures. The short-term, small-scale measures now being implemented are not always delivering the hoped-for results (Yamamoto 2006). For this reason the planning community needs spatially disaggregated Q_F data, at neighbourhood and city scales. Such information is practically impossible to derive by point *in-situ* fluxes measurements, while satellite remote sensing potentially is a valuable tool for estimating UEB parameters (e.g. albedo, surface cover) exploiting EO data. Currently estimation of Q_F spatial patterns is a challenge, but the expected near future satellite data may allow new methods to be developed to estimate the spatiotemporal patterns of all the UEB fluxes, including Q_F (Chrysoulakis *et al.* 2015). These techniques would allow benchmarking of current conditions so assessment of interventions can be made to ensure that the expected benefits are obtained. EO can directly support urban planning (Chrysoulakis *et al.* 2013a), map surface morphology and characteristics, and evaluate the implementation of UHI mitigation technologies (Mackey *et al.* 2012).

Systematic effort towards heat emissions reduction at local level is needed, as climate change mitigation policies may reduce CO_2 emissions at the point of energy production, but not necessarily heat emissions associated with the user. This implies that Q_F will continue to contribute to thermal discomfort and overheating (Smith *et al.* 2009), particularly in cities. Understanding Q_F spatiotemporal patterns is expected to lead to the development of more efficient methods and techniques for Q_F reduction in cities, supporting the mitigation of the UHI effect and improving thermal comfort.

Final comments

Anthropogenic heat emissions produced by human activities are one contributor to the UHI. Q_F sources include heat generated by the combustion process in vehicles, heat emitted by industrial processes, the conduction of heat through building walls, or emitted directly into the atmosphere by air-conditioning systems, and the metabolic heat produced by humans. The flux is strongly affected by climate and weather as this determines if heating or cooling systems are used. While Q_F varies spatially and temporally (diurnally, seasonally and yearly), under many conditions it can exceed energy receipt from net all-wave radiation. Its diurnal profile depends upon the day of the week and has local peaks in the morning and mid-afternoon, corresponding to peaks in transportation and building energy use. Q_F commonly is omitted from climate models despite its local importance (Allen *et al.* 2011). For this reason the Earth system modeling community needs spatially disaggregated Q_F data, at local and city scales. The estimation of Q_F spatiotemporal patterns is therefore a major challenge, which EO may be able to contribute to.

Knowledge of Q_F patterns is important for urban planning (e.g. to reduce or prevent Q_F hot spots), health (e.g. to estimate the impact on thermal comfort) and future proofing (e.g. to plan and implement interventions towards Q_F reduction in these areas). Planning tools, such as urban climatic maps and climatope maps, should be enriched with information on Q_F patterns. Mapping provides visualization of assessments of these phenomena to help planners, developers and policy makers make better decisions on mitigation and adaptation. The recently developed concept of LCZs (Stewart and Oke 2012) provides a means to better integrate Q_F in urban climate mapping.

Knowledge of the spatial and temporal patterns of Q_F are still prone to large uncertainties. Additionally a better understanding of the impact of Q_F on urban climate is needed as the impact overlaps at several scales: regional and meso-scale processes, plus within the street canyon and rooftop, and within the building. Better integration of the different estimation methods (*in-situ* measurements, modelling and remote sensing) would permit the respective strengths to be used and potential to overcome the weaknesses of individual methods. There is potential for EO to support our understanding of the role of Q_F within the UEB, but this remains underexploited. The H2020 project URBANFLUXES (www.urbanfluxes.eu) is expected to substantially contribute to this direction (Chrysoulakis *et al.* 2015). As Q_F is directly linked to urban metabolism, it was explored within the EU FP7 project BRIDGE (Allen *et al.* 2011, Iamarino *et al.* 2012, Kotthaus and Grimmond 2012, Chrysoulakis *et al.* 2013b, Kotthaus and Grimmond 2013, Lindberg *et al.* 2013). It is clear that the reduction of Q_F through interventions targeting urban metabolism optimization should be among the objectives of sustainable urban planning.

References

Allen, L., Lindberg, F. and Grimmond, C. S. B., 2011. Global to city scale urban anthropogenic heat flux: Model and variability. *International Journal of Climatology 31*, 1990–2005. doi: 10.1002/joc.2210.

Arnfield, A. J., 2003. Two decades of urban climate research: A review of turbulence, exchanges of energy and water, and the urban heat island. *International Journal of Climatology 23*, 1–16.

Bueno, B., Pigeon, G., Norford, L. K., Zibouche, K., and Marchadier, C., 2012. Development and evaluation of a building energy model integrated in the TEB scheme. *Geoscientific Model Development 5*, 433–448.

Christen, A., and Vogt, R., 2004. Energy and radiation balance of a central European city. *International Journal of Climatology 24*, 1395–1421.

Chrysoulakis, N., Esch, T., Gastellu-Etchegorry, J-P., Grimmond, C. S. B., Parlow, E., Lindberg, F., Del Frate, F., Klostermann, J., and Mitraka, Z., 2015. A novel approach for anthropogenic heat flux estimation from space. In Proceedings of the *36th International Symposium on Remote Sensing of Environment: Observing the Earth, Monitoring the Change, Sharing the Knowledge*, held in Berlin, Germany, May 11–15.

Chrysoulakis, N., Esch, T., Parlow, E., Düzgün, S. H., Tal, A., Sazova, A., Feigenwinter, C., Triantakonstantis, D., Marconcini, M. and Kavour, M., 2013a. The role of EO in sustainable urban planning and management: The GEOURBAN approach. In Proceeding of the *RSCy2013: First International Conference on Remote Sensing and Geoinformation of Environment*, held in Pafos, Cyprus, April 8–10.

Chrysoulakis, N., Lopes, M., San José, R., Grimmond, C. S. B., Jones, M. B., Magliulo, V., Klostermann, J. E. M., Synnefa, A., Mitraka, Z., Castro, E., González, A., Vogt, R., Vesala, T., Spano, D., Pigeon, G., Freer-Smith, P., Staszewski, T., Hodges, N., Mills, G., and Cartalis, C., 2013b. Sustainable urban metabolism as a link between bio-physical sciences and urban planning: The BRIDGE project. *Landscape Urban Planning 112*, 100–117.

Chrysoulakis, N., Vogt, R., Young, D., Grimmond, C. S. B., Spano, D., and Marras, S., 2009. ICT for urban metabolism: the case of BRIDGE. In: Wohlgemuth, V., Page, B., and

Voigt, K. (eds): *Proceedings of EnviroInfo2009: Environmental Informatics and Industrial Environmental Protection: Concepts, Methods and Tools Vol. 2* (pp. 183–193). Berlin: Hochschule für Technik und Wirtschaft Berlin.

Flanner, M. G., 2009. Integrating anthropogenic heat flux with global climate models. *Geophysical Research Letters 36*(2). doi:10.1029/2008GL036465.

Grimmond, C. S. B., 1992. The suburban energy balance: Methodological considerations and results for a mid-latitude west coast city under winter and spring conditions. *International Journal of Climatology 12*, 481–497. doi: 10.1002/joc.3370120506.

Grimmond C. S. B., Blackett, M., Best, M. J., Baik, J. J., *et al.*, 2011. Initial Results from Phase2 of the International Urban Energy Balance Comparison Project. *International Journal of Climatology 31*, 244–272.

Grimmond, C. S. B., Roth, M., Oke, T. R., Au, Y. C., Best, M., *et al.* 2010. Climate and more sustainable cities: Climate information for improved planning and management of cities (producers/capabilities perspective). *Procedia Environmental Science 1*, 247–274.

Hamilton, I. G., Davies, M., Steadman, P., Stone, A., Ridley, I., and Evans, S., 2009. The significance of the anthropogenic heat emissions of London's buildings: A comparison against captured shortwave solar radiation. *Building and Environment 44*, 807–817.

Heiple, S., and Sailor, D. J., 2008. Using building energy simulation and geospatial modeling techniques to determine high resolution building sector energy consumption profiles. *Energy and Buildings 40*, 1426–1436.

Iamarino, M., Beevers, S. and Grimmond, C. S. B., 2012. High-resolution (space, time) anthropogenic heat emissions: London 1970–2025. *International Journal of Climatology 32*, 1754–1767. doi: 10.1002/joc.2390.

Kato, S., and Yamaguchi, Y., 2005. Analysis of urban heat-island effect using ASTER and ETM+ Data: Separation of anthropogenic heat discharge and natural heat from sensible heat flux. *Remote Sensing of Environment 99*, 44–54.

Kato, S., and Yamaguchi, Y., 2007. Estimation of storage heat flux in an urban area using ASTER data. *Remote Sensing of Environment 110*, 1–17.

Kato, S., Yamaguchi, Y., Liu, C. C., and Sun, C. Y., 2008. Surface heat balance analysis of Tainan City on March 6, 2001 using ASTER and Formosat-2 data. *Sensors 8*, 6026–6044.

Kikegawa, Y., Genchi, Y., Yoshikado, H., and Kondo, H., 2003. Development of a numerical simulation system toward comprehensive assessments of urban warming counter-measures including their impacts upon the urban buildings' energy-demands. *Applied Energy 76*, 449–466.

Kotthaus, S., Grimmond, C.S.B., 2012. Identification of micro-scale anthropogenic CO_2, heat and moisture sources – processing eddy covariance fluxes for a dense urban environment. *Atmospheric Environment, 57*, 301–316.

Kotthaus, S., Grimmond, C.S.B., 2013. Energy exchange in a dense urban environment – Part I: Temporal variability of long-term observations in central London. *Urban Climate 10*, 261–280.

Lindberg, F., Grimmond, C. S. B., Yogeswaran, N., Kotthaus, S., and Allen, L., 2013. Impact of city changes and weather on anthropogenic heat flux in Europe 1995–2015. *Urban Climate 4*, 1–15.

Mackey, C. W., Lee, X., and Smith, R. B., 2012. Remotely sensing the cooling effects of city scale efforts to reduce urban heat island. *Building and Environment 49*, 348–358.

Mitraka, Z., Chrysoulakis, N., Del Frate, F., Gastellu-Etchegorry, J-P., 2015. Exploiting Earth Observation data products for mapping local climate zones. In Proceedings of *Joint Urban Remote Sensing Event JURSE 2015*, held in Lausanne, Switzerland, March 30–April 1.

Moriwaki, R., Kanda, M., Senoo, H., Hagishima, A., and Kinouchi, T., 2008. Anthropogenic water vapor emissions in Tokyo. *Water Resources Research 44*. doi:10.1029/2007 WR006624.

Narumi, D., Kondo, A., and Shimoda, Y., 2009. Effects of anthropogenic heat release upon the urban climate in a Japanese megacity. *Environmental Research 109*, 421–431.

Pigeon, G., Legain, D., Durand, P., and Masson, V., 2007. Anthropogenic heat release in an old European agglomeration (Toulouse, France). *International Journal of Climatology 27*, 1969–1981.

Ren, C., Ng, E. Y., and Katzschner, L., 2011. Urban climatic map studies: A review. *International Journal of Climatology 31*, 2213–2233.

Rigo, G., and Parlow, E., 2007. Modelling the ground heat flux of an urban area using remote sensing data. *Theoretical and Applied Climatology 90*, 185–199.

Roberts, S. M., Oke, T. R., Grimmond, C. S. B. and Voogt, J. A., 2006. Comparison of four methods to estimate urban heat storage. *Journal of Applied Meteorology and Climatology 45*, 1766–1781.

Sailor, D. J., 2011. A review of methods for estimating anthropogenic heat and moisture emissions in the urban environment. *International Journal of Climatology 31*, 189–299.

Sailor, D. J., Brooks, A., Hart, M., and Heiple, S., 2007. A bottom-up approach for estimating latent and sensible heat emissions from anthropogenic sources. In the Proceedings of the *American Meteorological Society: 7th Symposium on the Urban Environment*, held in San Diego, California, September 9–13.

Sailor, D. J., and Lu, L., 2004. A top-down methodology for developing diurnal and seasonal anthropogenic heating profiles for urban areas. *Atmospheric Environment 38*, 2737–2748.

Santamouris, M., 2014. Cooling the cities – A review of reflective and green roof mitigation technologies to fight heat island and improve comfort in urban environments. *Solar Energy 103*, 682–703.

Scherer, D., Fehrenbach, U., Beha, H-D., and Parlow, E., 1999. Improved concepts and methods in analysis and evaluation of the urban climate for optimizing urban climate processes. *Atmospheric Environment 33*, 4185–4193.

Seto, K. C., and Christensen, P., 2013. Remote sensing science to inform urban climate change mitigation strategies. *Urban Climate 3*, 1–6.

Smith, C., Lindley, S., and Levermore, G., 2009. Estimating spatial and temporal patterns of urban anthropogenic heat fluxes for UK cities: The case of Manchester. *Theoretical and Applied Climatology 98*, 19–35.

Stewart, I. D., and Oke, T. R., 2012. Local climate zones for urban temperature studies. *Bulletin of the American Meteorological Society 93*, 1879–1900

Stone B., 2009. Land use as climate change mitigation. *Environmental Science and Technology 43*, 9052–9056.

Stone, B., Vargo, J., and Habeeb, D., 2012. Managing climate change in cities: Will climate action plans work? *Landscape Urban Planning 107*, 263–271.

United Nations, 2010. *United Nations world population prospects: The 2010 revision and World urbanization prospects: The 2011 revision*. Population Division of the Department of Economic and Social Affairs of the United Nations Secretariat. Available at: http://esa.un.org/unup/unup/ index_panel1.html.

Weng, Q., 2009. Thermal infrared remote sensing for urban climate and environmental studies: Methods, applications, and trends. *ISPRS Journal of Photogrammetry and Remote Sensing 64*, 335–344.

Xu, W., Wooster, M. J., and Grimmond, C. S. B., 2008. Modelling of urban sensible heat flux at multiple spatial scales: A demonstration using airborne hyperspectral imagery of Shanghai and a temperature–emissivity separation approach. *Remote Sensing of Environment 112*, 3493–3510.

Yamamoto, Y., 2006. Measures to mitigate urban heat islands. *Quarterly Review Science & Technology Trends 18*, 65–83.

3

VALUING GREEN SPACES AS A HEAT MITIGATION TECHNIQUE

Steve Kardinal Jusuf[1] *and Wong Nyuk Hien*[2]

[1] Singapore Institute of Technology – Singapore
[2] National University of Singapore – Singapore

Urbanization and greenery

According to the United Nations (2012), the world urban population has grown exponentially from 0.75 billion in 1950 to 3.6 billion in 2011. It is expected to further increase by 72 per cent to 6.3 billion by 2050. The accelerated growth of urbanization and industrialization has resulted in extensive modifications to the natural land cover and affected the absorption of heat radiation and energy balance (Oke, 1973; Parham and Fariborz, 2010). The high concentration of artificial impervious surfaces and built structures, further aggravated by anthropogenic heat emissions, has led to the 'urban heat island' (UHI) phenomenon, whereby air temperatures in densely built urban areas are higher than the surrounding non-urban areas (Kolokotroni, *et al.*, 2006).

The relationship between urban development and greenery is mostly negative. According to Khadpecker and Jacob (2004), 'shrubs, grasses, trees and other forms of natural vegetation are usually the first victims of urbanization' (p. 189). Buildings and vegetation are competing for space in the city. The diminishing of greenery has led to the increase of air temperature, i.e. UHI effect.

As shown in Figure 3.1, the thermal satellite image of Singapore's urban heat island study shows that the high dense built-up areas mainly in the southern part of Singapore are warmer and represented in yellow to orange colors, while less-developed areas in the northern part of Singapore are much cooler and represented as green colors. A closer look at the satellite image and the thermal satellite images shows that the temperature distributions are closely related with the distribution of buildings and vegetation. For example in Figure 3.2, the images show part of Singapore's residential area. The residential area usually has open spaces of greenery for community parks and sports fields. Thus, there are areas of yellowish and green colors in between red-color areas, which show that the greenery areas have favorable impact to the surrounding buildings.

Various forms of greenery exist in the city areas, such as nature reserves, parks, rooftop gardens, vertical greeneries, but they are mainly categorized into two major categories: natural and man-made, as shown in Figure 3.3. Natural landscapes of the aboriginal flora, fauna or geological features preserved in the built environment are called natural reserves. However, as cities are developing, the existence of natural reserves is diminishing. The majority of vegetation in the city which is designed, planted and maintained belongs to man-made landscape. City parks usually have larger areas as compared to neighborhood

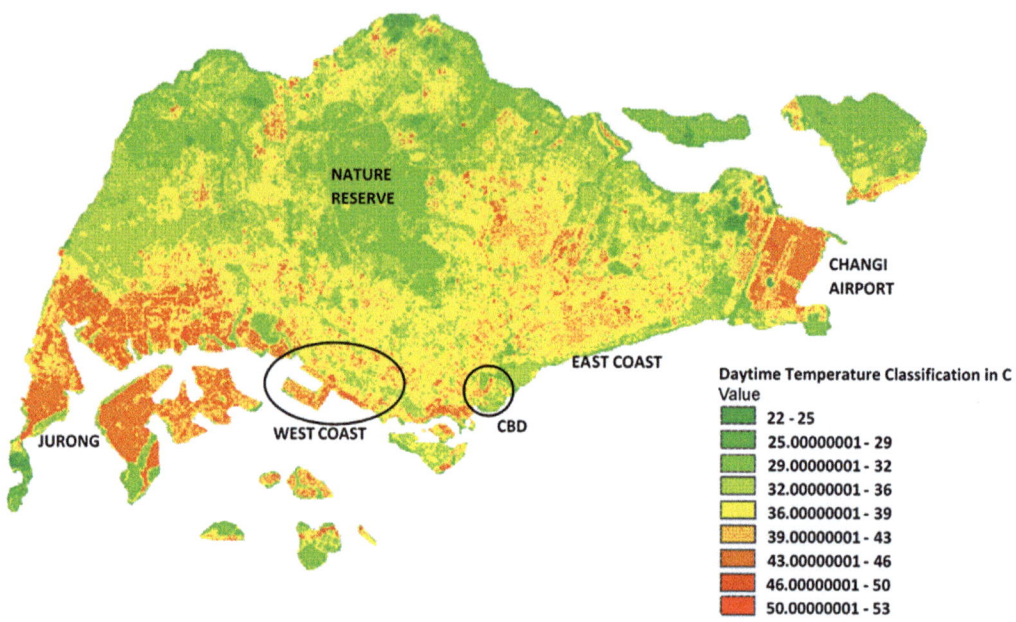

Daytime Temperature Classification in C
Value

	22 - 25
	25.00000001 - 29
	29.00000001 - 32
	32.00000001 - 36
	36.00000001 - 39
	39.00000001 - 43
	43.00000001 - 46
	46.00000001 - 50
	50.00000001 - 53

Figure 3.1
Singapore urban heat island study - thermal satellite image.
Source: Jusuf, et al., 2007. Reprinted from Habitat International, 38 /1, Steve Kardinal Jusuf, NH. Wong, Emlyn Hagen, Roni Anggoro and Yan Hong, The influence of land use on the urban heat island in Singapore, 236, 2007, with permission from Elsevier.

parks. Building these parks has various purposes, including preserving natural landscape, providing recreational facility and beautifying the built environment. Other green areas Include the vegetation in between urban spaces, such as roadside planting and vegetation in spaces between buildings.

In the last decade, landscape in buildings has gained popularity as an effort to increase greenery areas in high-density cities, especially rooftop garden and vertical greenery. Rooftop gardens can be categorized into two types: intensive and extensive, as shown in Figure 3.4. Intensive rooftop gardens require relatively thick substrate for various plant types to grow, including grass and trees. They are usually designed to be accessible for people and are used as parks or building amenities. The extensive one features lightweight growing media for grass or turf.

Although introducing plants on the building façades is not a new concept, vertical greenery becomes a common feature in buildings which are designed to be 'green'. More and more research is conducted to develop better irrigation systems, growing media and construction systems. Vertical greenery can be simply divided into three fundamental types: wall-climbing, hanging-down and the most recent one, module type. The wall-climbing type is a very traditional example of the vertical climbing method. The plants can either cover the wall of buildings naturally or grow upwards with the help of a supporting system. The hanging-down type is also a popular method of vertical landscaping, which can form a vertical green belt in multi-storey buildings. The module type is the latest vertical greenery system that requires a proper design irrigation system, structured growing media and a selection of plants (Wong and Chen, 2009). Figure 3.5 shows some examples of vertical greenery on buildings' façades.

Figure 3.2
Part of Singapore's superimposed satellite and thermal satellite images.
Source: Google map and Landsat-7 ETM+ image of Singapore acquired on 11 October 2002.

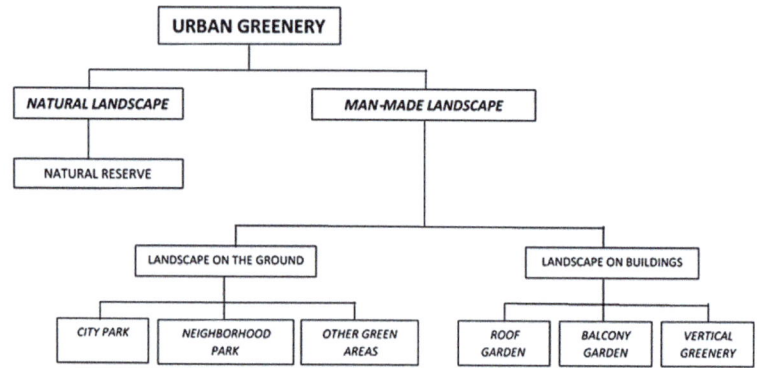

Figure 3.3 Formations of urban green.
Source: Adapted from Wong and Chen, 2009.

Figure 3.4 Rooftop garden: intensive system (top) and extensive system (below).
Photos by Jusuf.

The benefits of greenery have been investigated extensively by many researchers from thermal aspects to social aspects. From the thermal aspect, the presence of greenery in urban areas helps to cool the environment through the process of evapotranspiration, where large amount of solar radiation can be transformed into latent heat, converting water from liquid to gas which in turn results in a lower leaf temperature, lower ambient air temperature and higher humidity (Chen and Wong, 2006). It can reduce long-wave radiation exchange between buildings due to the low surface temperatures created by shading of plants (Wong, *et al.*, 2003), as the leaves of deciduous trees can intercept from 75 per cent to 90 per cent of incoming solar radiation on a clear day (Heisler, 1986). With reduction of incoming solar radiation on buildings, there is a decrease in cooling loads, but there may be an increase in heating loads in winter (Canton, *et al.*, 2001). Evapotranspiration contributes to creating lower-temperature spaces in an urban environment, known as 'the oasis phenomenon' (Santamouris, 2001), which is characterized by Bowen ratios in the vegetative canopies of 0.5– 2 (Taha, 1997). A proper arrangement of plants around buildings improves both psychological effects and unfavorable microclimatic conditions (Robinette, 1972).

In this chapter, several types of urban greenery are discussed, including urban parks, rooftop gardens and vertical landscaping.

Figure 3.5
Vertical greenery system.
Photos by Jusuf.

Urban parks

Urban parks have great roles in the cities in providing natural spaces for city dwellers to meet their psychological needs and social functions, as key ingredients for a city's sustainability. People have positive feelings and less stress as experiencing nature in urban environment enhances contemplativeness, rejuvenates the city dwellers and provides sense of peacefulness and tranquility (Ulrich, 1981; Kaplan 1983; Hartig, *et al.*, 1991; Schroeder, 1991; Conway, 2000; Chiesura, 2004). A number of studies show positive correlations between large green space areas and lower stress regardless of the intensity of the activities inside the parks (Barton and Pretty, 2010; Thompson, *et al.*, 2012). Adequate provision of green space becomes an important message for urban planners and urban designers in designing new developments or urban revitalization.

Urban parks are one of the main strategies to mitigate the UHI effect, as greenery has an important role in modifying urban climate (Jauregui, 1990/1991;

Figure 3.6
**The comparison of average
air temperature measured
at different locations in
Bukit Batok Park**
Source: Chen and Wong,
2006. Reprinted from Energy
and Buildings, 38 /2, Chen
Yu, Wong Nyuk Hien,
Thermal benefits of city
parks, 112, 2006, with
permission from Elsevier

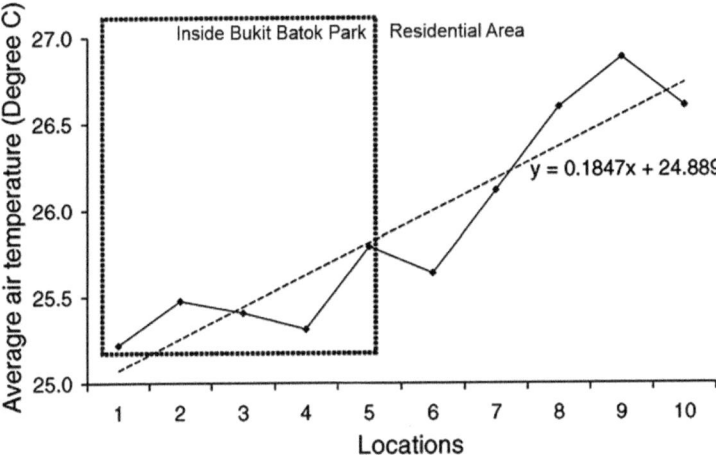

Kawashima, 1990/1991; Ca, et al., 1998). The thermal impacts of urban parks can be observed not only inside the parks but also the surrounding built environment. A study in tropical climate in Singapore investigated the microclimate condition of a neighborhood park (i.e. Bukit Batok Park) and its influence to the surrounding environment (Chen and Wong, 2006). Temperature sensors were installed in the middle of the park and correspondingly moved away from the park to the surrounding residential area. As shown in Figure 3.6, the closer to the parks, the lower the temperature is. Average temperature of about 1.3°C difference was observed at locations around the parks. Meanwhile, in hot and dry climate of Lisbon, Portugal, the urban park was found to be cooler both in the sunny and in the shaded areas. During clear and calm weather, the thermal differences between the park and its surroundings were much higher. In the hottest day, the maximum and minimum temperatures were 7.8°C and 8.3°C above the average, respectively (Oliveira, et al., 2011). The cooling effect of urban parks on surrounding areas is strongly related to several factors including distance to the parks, wind direction and building layout.

During daytime, the rate of cooling is controlled by shading, surface albedo and water availability, while thermal properties of surfaces and their radiative geometry are the most important controls at night. In the afternoon, irrigated green spaces (such as garden, multi-use and savannah type parks) are cooler than dry grass parks, but the surface of parks with extensive tree coverage are warmer than open parks (with higher sky view factors and dry soils) at night (Spronken–Smith and Oke, 1998). Plants may prevent the outgoing thermal radiation to the sky from a surface and the ventilation in a space covered, resulting in higher air and globe temperature in the early evening, but overall, space and surface covered or shaded by vegetation show lower temperatures than those not shaded (Hoyano, 1988). Another greenery research in the tropical country of Singapore concluded that trees reduce the sky openness, i.e. sky view factor (SVF), in the built environment. During daytime, there is a positive correlation between air temperature and SVF; the higher the SVF, the higher air temperature is. While during nighttime (after midnight), the measurement showed a statistically insignificant negative correlation between air temperature and SVF, i.e. trees do not trap the heat released to the sky (Wong and Jusuf, 2010).

Inside the parks, several studies investigated the thermal sensation experienced by their visitors. The thermal sensation varies in different zones of the park depending on factors, such as vegetation density, sky view factor, wind speed and surface albedo. Hoyano (1988) found that air temperature of up to about 1 m from the ground surface is lower under a planted pergola compared to open ground. Introducing grass under trees further reduces the ground temperature, thus contributing even more to human comfort (Shashua–Bar, *et al.*, 2011). The rate of cooling also depends on the vegetation and irrigation regime, the adjacent urban fabric, and the aridity of the location (Shashua–Bar, *et al.*, 2009). In observation of surface energy balance in an irrigated urban park, strong advective effects on evaporation are driven by the differences in surface and air temperature, and humidity, between the cool, wet park and its warmer, drier built-up surroundings (Spronken–Smith, *et al.*, 2000). To create a thermally comfortable park, not only minimizing the area of hardscape pavement and careful design of shading plants are encouraged, but also creation of large bodies of water are desired to maintain thermal comfort during hot and cold months (Streiling and Matzarakis, 2003; Mahmoud, 2011).

A study in desert climate found that trees in liman (floodwater irrigated plot) provide shading which reduced radiation, evaporation rate and wind speed if compared to open desert. Based on black-body temperature analysis, the liman provides greater thermal comfort during summer, but is cooler than open desert during winter (Schiller and Karschon, 1974).

The radiant exchange, particularly in deserts, provides thermal comfort not only by directly shading, but also by reducing long-wave emission from surfaces and by limiting the amount of solar radiation reflected. It was found that irrigated grass has high water demand, and this demand can be much lowered by shading the grass. In terms of thermal comfort, shaded grass provides better comfort than shaded paving, followed by unshaded grass and unshaded paving as the worst (Shashua–Bar, *et al.*, 2011).

This temperature difference created by urban parks on the surrounding areas may lead to savings of cooling energy consumption from 9 per cent up to as high as 50 per cent for strategic landscaping (Meier, 1990/1991; Wong, *et al.*, 2011) and improvement of outdoor thermal comfort for city dwellers (McPherson, *et al.*, 1988).

Vegetation can reduce the cooling loads of buildings by modifying air temperature, solar heat gain, long-wave heat gain, and convection heat gain through shade, vegetation surface temperatures and emissivity, and evapotranspiration (Hutchison, *et al.*, 1982; McPherson, *et al.*, 1989). Huang, *et al.* (1987) modelled the effects of landscaping on temperature, humidity, wind speed and solar gain for residential buildings in four cities. Using evapotranspiration models for arid and semi-arid climates, it was found that most of the cooling energy savings are due to the effect of evapotranspiration and only 10 per cent to 30 per cent are due to shading. Shading and evaporative cooling of the air surrounding the condenser lowers supply air temperature by about 4°C, which improves the unit's coefficient of performance (COP) by as much as 10 per cent (Parker, 1983).

An analysis for outdoor cooling in a hot-arid region considering the efficiency of water use in summer found that shaded grass produced greater cooling as well as a reduction of more than 50 per cent in total water use compared to unshaded grass (Shashua–Bar, *et al.*, 2009).

The selection of vegetation species is also important in influencing the outdoor thermal comfort as different species have different shading and evapotranspiration capabilities. In the Mediterranean climate of Greece, the *Ficus* genus (a tropical plant) was found to have the highest amount of evaporation followed by fig, pine, palm, bitter orange and olive respectively (Georgi and Dimitriou, 2010).

Urban parks have positive role in modifying not only urban microclimate but also urban air pollution, which has become a major problem in big cities worldwide. A number of studies have looked into the ability of vegetation to improve local air quality (e.g. Fujii, *et al.*, 2005; Freiman, *et al.*, 2006; Nowak, 2006; Nowak, *et al.*, 2006; McDonald, *et al.*, 2007; Yin, *et al.*, 2007). The ability of vegetation to reduce the ambient temperature can slow down photochemical reactions and lead to less secondary pollutants such as ozone (Rosenfeld, *et al.*, 1998; Akbari, 2002). Trees are able to intercept various atmospheric particles and absorb various gaseous pollutants (Bealey, *et al.*, 2007). A study in Aachen, Germany, found lower nitrogen oxide (NO) and nitrogen dioxide (NO_2) concentrations inside open green spaces, where wind speed and direction play an important role, as compared to the locations closer to main roads (Kuttler and Strassburger, 1999). Similar results were also found on field measurement studies in Hong Kong (Lam, *et al.*, 2005) and Tel Aviv, Israel (Cohen and Potchter, 2010). Makhelouf (2009) studied the role of green spaces on urban climate and air pollution (i.e. carbon monoxide (CO) and sulphur dioxide (SO_2)) at different types of gardens in Paris over a period of four years. Pollutants penetrate the public gardens easily, which are less than 1 hectare from the surrounding urban environment. However, the planted areas remain less polluted than nearby built-up areas. Large planted areas such as parklands are cooler than the surrounding areas, driving the pollutants with the air movement out of the planted areas. The significant number of trees, being the source of the breeze, plays an important role in pushing back the pollutants. Hence, large parklands are less polluted than the gardens (surface lower than 100 hectares) in all types of weather.

Trees in urban street canyons may reduce wind speeds and, hence, reduce traffic pollutant dispersion. Tree layout should be planned based on the approaching wind direction, the street-canyon aspect ratio, the crown porosity and the tree-stand density (Gromke and Ruck, 2012).

Although some vegetation emits higher hydrocarbons that play a role in the formation of photochemical smog, the impact can be reduced by careful choice of species (Corchnoy, *et al.*, 1992; Benjamin, *et al.*, 1996) and layout planning (Grossi, 2008). Greenery decreases photochemical reactions that form pollutants such as ozone in the atmosphere by lowering temperatures (Rowe, 2011). 'Reduce car emissions, and cities won't have to worry about the trees' (Fischetti, 2014).

Rooftop garden

The rooftop garden is considered as one of the effective methods in mitigating UHI effect (Gomez, *et al.*, 1998; Takebayashi and Moriyama, 2007; Alexandri and Jones, 2008) including outdoor thermal condition, building energy consumption, urban air pollution and storm water runoff.

The reduction of cooling energy of a building may be the result of the insulation potential of the roof covered with vegetation rather than the low temperature of the external surface (Santamouris, 2014; Barrio, 1998). The rate of cooling depends on the leaf area index (LAI), foliage geometrical characteristics, soil apparent density, thickness and moisture content (Barrio, 1998). However, for similar indoor conditions and U-values of the roof, the technology that decreases the cooling consumption the most presents a lower temperature in the external surface, a higher sensible heat flux reduction and very probably a higher mitigation potential (Santamouris, 2014).

A study in Malaysia (Rashid and Ahmed, 2009) found that the maximum difference of temperature between bare roof surface and wet soil surface was 22°C. The temperatures of the green roof stay within a low temperature range due to the evaporative cooling effect of the green roof. The green roof reduces the surface temperature on the ceiling under the roof slab by a maximum of 3.0°C and an average of 1.7°C compared to bare concrete roof. With the use of green roof, the daily average indoor air temperature can be reduced by 3.0°C, i.e. from 33.0°C to 30.0°C. The maximum temperature difference between indoors and outdoors was 4.4°C with green roof and only 2.8°C for bare concrete roof.

The surface temperature of a bare concrete roof surface can be up to 57°C in the early afternoon with a 20°C surface temperature fluctuation for the whole day, while a rooftop surface with shrubs has only a maximum surface temperature of 26.5°C with 3°C daily variation. The lower the surface temperature the lower heat transferred into building's indoor environment. The total daily heat gain of bare roof concrete was calculated to be 742.9 kJ as compared to 164.3 kJ for the rooftop surface with shrubs, a reduction of 78 per cent (Wong and Chen, 2009).

Green roofs have the benefit not only in reducing heat gain but also in reducing heat loss through providing more insulation for the building during winter. Green roofs reduced heat gains by 70 per cent and 83 per cent during warm periods as compared to ceramic and metallic roofs, respectively, and heat losses by 44 per cent and 52 per cent during cold periods as compared to ceramic and metallic roofs, respectively (Parizotto and Lamberts, 2011; D'Orazio, et al., 2012). It means green roofs help reduce annual cooling energy consumption between 2 per cent and 49 per cent and reduce annual heating energy consumption between 8 per cent and 46 per cent (Niachou, et al., 2001; Santamouris, et al., 2007; Jaffal, et al., 2012).

Extensive green roofs are a proven technology in temperate or tropical countries, but there are many barriers to the implementation of such roofs in Mediterranean countries with year-round or seasonal hot, dry climates. Extensive planted roofs may lower indoor air temperature in summer during the warmest daytime hours, but this effect is due mostly to the shading effect of the plant canopy, which reduces the penetration of radiant energy to the soil. The thermal perform-ance of an extensive planted roof in winter may be little better, especially if the soil is moist or if evaporation occurs at the surface. Wind increases the water consumption of the plant cover. The reduction in the sensible heat load of the model building attributed to the green roof system was less than 5 per cent of the latent heat content of the water lost to evapotranspiration (Schweitzer and Erell, 2014).

Studying the effect of the rooftop garden on the street level, simulation studies were carried out. In New York, extensive roofs covered with grass can decrease the daily average temperature by 0.3–0.55 K (Savio, et al., 2006). In Tokyo, the potential of vegetative roofs in medium- and high-rise buildings to decrease ambient temperature is almost negligible (Chen, et al., 2009). In Hong Kong, it was found that when the building height to street width (aspect ratio) exceeds 1, the possible cooling benefits are low (Ng et al., 2012). Experimental study in Singapore also found that the cooling effect of vegetative roofs on pedestrian level may be effective when the building height is lower than 10 m (Wong et al., 2003). In Taipei, it was measured that green roofs decrease the ambient air temperature by 0.26°C in average and the maximum temperature by 1.6 K (Sun, et al., 2012).

In terms of heat mitigation, recently there have been studies comparing green roofs and cool roofs. As summarized by Santamouris (2014), the differences in performance depend on the climate, the albedo of the cool roofs, the LAI and the soil moisture content of the green roofs. Cool roofs present higher heat island mitigation potential when the albedo is equal or higher than 0.7. Otherwise, high LAI and well-irrigated green roofs perform better, particularly in drier climates. A study in New York City found that a green roof can further increase energy savings by 40–110 per cent compared to a white roof (Susca, et al., 2011). Although green roofs are more expensive to install than white roofs, other benefits besides temperature reduction include storm water runoff mitigation, roof-service lifetime extension, building-amenity value and biodiversity value. White roofs require high maintenance for the performance to be fully realized (Gaffin, 2006).

Modelling studies were also conducted to understand the ability of green roofs in reducing air pollutants. Green roofs were predicted to remove 58 metric tons of air pollutants if all roofs in Washington, DC, were converted into green roofs (Deutsch, et al., 2005). By covering 20 per cent of the roof surface in Chicago, the reduction of NO_2 was between 806.48 and 2769.89 metric tons depending on the type of plants used (Corrie, et al., 2005). Another study calculated that covering all the roofs with 100 per cent green roofs would remove 1405.50 and 2046.89 metric tons if extensive and intensive roofs were used, respectively (Yang, et al., 2008), although the uptake of NO_2 is influenced by several factors including meteorological conditions, concentration of NO_2 and plant physiology. In a sunny day, a green roof may lower the carbon dioxide (CO_2) concentration in the nearby region as much as 2 per cent, and the CO_2 concentration above the green roof is measured to be 4.3 mg/m³ lower than above bare roof (Li, et al., 2010). A field study on a 4000 m² green roof in Singapore, Tan and Sia (2005) found that the levels of particles and SO_2 in the air above the roof were reduced by 6 per cent and 37 per cent respectively after installation of the green roof.

Another benefit of green roofs is their role in rainfall and storm water management. A number of studies reported significant reduction of water runoff generation. Rainwater retention of green roofs ranges from 45 per cent to 100 per cent of incoming rainfall (Liesecke, 1998; Schade, 2000; Monterusso, et al., 2004; Mentens, et al., 2006). A moderate application of extensive green roofs with 100 mm substrate depth on 10 per cent of the buildings in Belgium was calculated to reduce rainwater runoff by 2.7 per cent for the region (Mentens et al.., 2006). In the Mediterranean region, extensive green roof has an average retained volume and peak reduction equal to 68 per cent ± 37 per cent and 89 per cent ± 15 per cent respectively (Fioretti, et al., 2010). Plants with leaves can retain water even better than bare soil (Hoyano, 1988).

The reduction of water runoff is mainly attributed to several factors. Green-roof systems absorb rainwater, delaying the initial time of runoff. Part of the rainwater is also retained by the system, hence reducing the total runoff and distributing the runoff through a relatively slow release of the excess water which is temporarily stored in the pores of the substrate (Moran, *et al.*, 2003; Villarreal, *et al.*, 2004; Mentens *et al..*, 2006). The performance of green roof in reducing water runoff is determined by slope of the roof and characteristics of substrate including types and depths of the substrate (Monterusso *et al..*, 2004; VanWoert, *et al.*, 2005; Berndtsson, 2010). A steeper roof slope retains less rainwater, while thicker substrate retains more rainwater in the green-roof system. In a recent study, plant species was also found as another important factor. Grass is the most effective plant in reducing water runoff followed by forbs and sedum. Grass and forbs with taller height, larger diameter, and larger shoot and root biomass are preferred, as their structural properties promote better rainwater interception (Lundholm, *et al.*, 2010; Nagase and Dunnett, 2012).

The criteria of vegetation for green roof include drought tolerance, solar-radiation tolerance and the cooling ability of plants. There are still ongoing debates on whether to use native or non-native plants (Li and Yeung, 2014).

Vertical landscaping

Strategically planted vegetation on high-rise building façades is relatively new as compared to rooftop greenery. However, in recent years, it has become commonly recognized as a method to increase the amount of greenery in the high-dense built environment due to rapid research and technology development of vertical greenery systems.

Figure 3.7
The modular trellis panel system (left) and its close up (right)
Photos by Jusuf.

Figure 3.8
The cable and wire-rope
net system (left) and its
close up (right).
Photos by Jusuf.

Vertical greenery systems can be broadly classified into two major categories: green façade and living wall. Climbing plants or cascading groundcovers trained to cover specially designed supporting structures are known as green façade systems (GRHC, 2008). Two green façade systems that are frequently used are the modular trellis panel as well as the cable and wire-rope net systems.

The modular trellis panel system consists of rigid, lightweight and three-dimensional panels made of steel wire that support plants with both a face grid and a panel depth. The concept is to keep the vegetation at a certain distance away from the wall, as shown in Figure 3.7. The cable and wire-rope net system as seen in Figure 3.8 uses both cables and a wire net, allowing for greater flexibility and degree of design applications.

Living wall systems consist of pre-vegetated panels that are fixed vertically to a structural wall. These panels can be made of a wide range of materials including plastic, synthetic fabric, clay or metal, and support a great diversity and density of plants. These systems give a totally different feel, as the plants are not climbers but small, creeping herbaceous perennials, ferns, grasses and small shrubs. However, an irrigation system needs to be built into the structure and is therefore more costly and complicated to install.

A variety of alternative designs have been derived from the living wall systems, adopting more sophisticated designs which include the use of planter-cassettes or modular system with patented substrate as shown in Figure 3.9 and Figure 3.10, respectively.

Figure 3.9
The plant-cassette system (left) and its close up (right).
Photos by Jusuf

Throughout history, vertical greenery systems have been used to control indoor climates or to provide visual and aesthetics purposes. Both rooftop gardens and vertical greenery systems share similar quantitative benefits. The increase in the concentration of built structures with heat-absorbing properties, the reduction of evaporating surfaces and the lack of vegetation cover lead to the UHI effect. Figure 3.11 illustrates the temperature profile variations in summer and winter with and without vertical greenery systems (City of Stuttgart, 2008). Vegetation through their cooling effect and the creation of complex air flows can minimize the convection currents on the side of a solar heated building and dust generation (Dunnett and Kingsbury, 2008).

A number of studies have quantified the thermal benefits of covering the building envelope with vegetation on the microclimate of built environment for various climates and urban canyon geometries. In humid climates of Hong Kong, vertical greenery systems can achieve substantial benefits by temperature decrease up to 8.4°C, as they have a stronger effect than rooftop gardens in decreasing the air temperature in a canyon (Alexandri and Jones, 2008). On hot sunny days in Chicago, a plant layer on a brick façade was estimated to reduce its exterior surface temperature by 0.7–13.1°C, reducing the heat flux through the exterior wall by 2–33 W/m², and providing an effective R-value of 0.0–0.71 m²K/W, depending primarily on wall orientation, leaf area index, and radiation attenuation coefficient (Susorova, *et al.*, 2013). The unshaded building's wall

Figure 3.10
The modular system (left) and its close up (right).
Photos by Jusuf

surface temperature in Mediterranean continental climate was on average approximately 5.5 °C higher than in areas partially covered by vegetation. This difference was higher in August and September, reaching maximum values of 15.2 °C on the southwest side in September (Perez, *et al.*, 2011). Meanwhile, a study of various vertical greenery systems in the tropical climate of Singapore found that the temperature on the substrate surface was lower compared to the wall surface in the evening and night but with a reversal in the day. This was explained by the higher temperature of the substrate surface due to direct exposure to solar radiation in the day whilst the wall surface was covered by the plant panels, substrate and plants, resulting in a lower temperature. At night, the substrate surface with its high heat capacity tends to lose heat faster than the wall surface, which was covered behind the substrate and tends to retain heat, thus contributing to the higher temperature of the wall surface as compared to the substrate surface. The maximum temperature reduction of the wall and substrate surfaces can be more than 10°C. Meanwhile, the effects of vertical greenery systems on ambient temperature depend on specific vertical greenery systems. These results point to the potential thermal benefits of vertical greenery systems in reducing the surface temperature of building façades and ambient temperature close to the building façade in the tropical climate, leading to a reduction in the cooling load and energy cost (Wong, *et al.*, 2010). The temperature reduction inside a street canyon is significant because the distribution of ambient air in a canyon influences the energy consumption

WITHOUT GREENERY

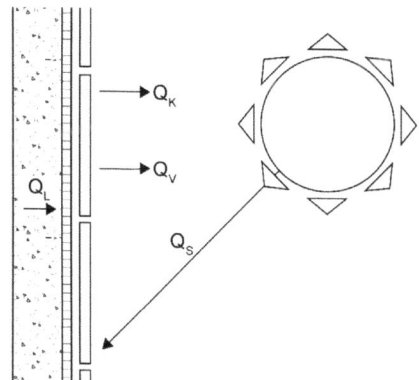

WITH GREENERY

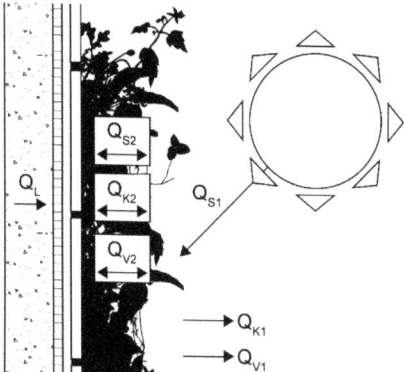

TEMPERATURE

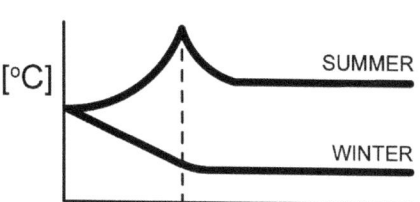

of buildings as higher temperatures in a canyon increase heat convection to a building, and correspondingly increases the cooling load (Niachou *et al.*., 2001).

Vertical greenery systems have better thermal insulation properties as compared to bare building façades, as they reduce the conduction heat gain through its shading (Wong, *et al.*, 2009) and especially for temperate climate, they also act as wind barrier (Perini, *et al.*, 2011). A façade fully covered by greenery is protected from intense solar radiation in summer and can reflect or absorb in its leaf cover between 40 per cent and 80 per cent of the received radiation, depending on the amount and type of greenery (City of Stuttgart, 2008). The shading effect of vertical greenery systems reduces the energy used for cooling by approximately 23 per cent and the energy used by fans by 20 per cent, resulting in an 8 per cent reduction in annual energy consumption (Bass and Baskaran, 2003), while using building energy simulation TAS software, the use of vertical greenery was found to be able to reduce the ETTV of a full glass-façade building by up to 40.68 per cent and reduce the building cooling energy consumption of up to 74.29 per cent in a non-windows hypothetical building model (Wong *et al.*, 2009). In warm temperate and arid climates, vertical greenery systems can reduce cooling energy between 5 per cent and 50 per cent (Perez, *et al.*, 2014).

Besides thermal benefits, vertical greenery systems offer other benefits, including improvement of water retention (Schmidt, 2009; Dunnett and Kingsbury, 2008). Vertical greenery systems, like rooftop gardens, when covered with vegetation do play some role in intercepting and temporarily holding water

Figure 3.11
Temperature profile for summer and winter climates with respect to wall with (left) and without (right) vertical greenery system.
Source: Adapted from City of Stuttgart, 2008

during rainstorms. The ability of vertical greenery systems to retain storm water varies by design. A vine-covered wall delays the runoff or allows a slow infiltration into a permeable ground cover. A window-shade design that involves plants in soil has similar benefits to a rooftop garden, where water is stored in the soil and later used by the plants or returned to the atmosphere by evapotranspiration.

Exposed building materials are subjected to large annual and diurnal fluctuations in temperature which cause expansion and contraction, leading to eventual failure. In addition, ultraviolet light causes wall materials to degrade and lose their strength. Soil and vegetation may protect walls from these threats, despite a widespread belief that vegetation threatens building structure. Although some species of plant can become rooted in mortar and damage masonry over a long period of time, most plants have little or no negative impact on buildings and actually protect the building fabric from the weather (Johnston and Newton, 1993).

Climbers on buildings can also help to protect the building surfaces from damage due to very heavy rainfall and play some role in intercepting and temporarily holding water during rainstorms. Protection of the wall-cladding results in a longer material lifespan. This helps to decrease the need for maintenance, which in turn increases associated savings from replacement costs.

Urban planning and assessment tool for greenery implementation: Case study of Singapore

Greenery clearly has a positive thermal impact on the built environment and may be the most effective strategy for mitigating the UHI effect. A number of studies relate the urban air temperature condition with the built environment (Bottyan and Unger, 2003; Giridharan, et al., 2007; Robinson and Bruse, 2011; Jusuf and Wong, 2012). At microscale, buildings and vegetation influences the incident solar radiation received by urban surface. Different urban morphology conditions create different urban microclimate conditions.

In Singapore, empirical models, named as Screening Tool for Estate Environment Evaluation (STEVE) Tool, correlate maximum temperature (T_{max}), average temperature (T_{avg}) and minimum temperature (T_{min}) of a point of interest with a radius of 50 meters in an estate with the urban morphology parameters (Wong and Jusuf, 2013). The models are as shown in Equations 1–5:

$$T_{min} \,(°C) = 2.116 + 0.932\,T_{min\text{-}r}\,(°C) - 0.272\,GnPR + 0.003\,PAVE\,(\%) + 1.619E\text{-}05\,WALL\,(m^2) - 0.004\,HBDG - 1.33\,ALB \tag{1}$$

$$T_{avg}\,(°C) = 2.676 + 0.919\,T_{avg\text{-}r}\,(°C) - 0.201\,GnPR - 0.015\,HBDG + 1.520E\text{-}05\,WALL\,(m^2) + 0.240\,SVF \tag{2}$$

$$T_{max}\,(°C) = 4.494 + 0.695\,T_{max\text{-}r}\,(°C) + 0.001\,SOLAR_{max}\,(W/m^2) - 0.39\,WIND_{max}\,(m/s) + 0.008\,PAVE\,(\%) - 0.002\,H_{avg}\,(m) + 1.078\,SVF + 32.235\,ALB \tag{3}$$

$$T_{avg(daytime)}\,(°C) = 7.619 + 0.733\,T_{avg(daytime)\text{-}r}\,(°C) + 3.61E\text{-}5\,SOLAR_{total}\,(W/m^2) + 0.004\,PAVE\,(\%) + 0.669\,SVF - 0.013\,HBDG \tag{4}$$

$$T_{avg(nighttime)}\,(°C) = 2.347 + 0.904\,T_{avg(nighttime)\text{-}r}\,(°C) + 5.786E\text{-}05\,SOLAR_{total}\,(W/m^2) + 0.007\,PAVE\,(\%) - 0.06\,GnPR - 0.015\,HBDG + 1.311E\text{-}05\,WALL\,(m^2) + 0.633\,SVF \tag{5}$$

The parameters in the above equations can be categorized into two groups: *climate* and *urban morphology* parameters. The *climate parameters* are daily minimum (T_{min-r}), average (T_{avg-r}), maximum (T_{max-r}), daytime average ($T_{avg(daytime)-r}$) and nighttime average ($T_{avg(nighttime)-r}$) temperatures at meteorological station, usually located outside the city. For the SOLAR parameters, the average of daily total solar radiation ($SOLAR_{total}$) is used in the $T_{avg(daytime)}$ and the $T_{avg(nighttime)}$ models, while the average of solar radiation maximum of the day ($SOLAR_{max}$) is used in the T_{max} model. Wind condition ($WIND_{max}$) is also found to influence the maximum temperature.

The *urban morphology parameters* are Green Plot Ratio (GnPR), percentage of pavement area (PAVE), average building height to building area ratio (HBDG), average building height (H_{avg}), total wall surface area (WALL), sky view factor (SVF) and average surface albedo (ALB). Greenery is one of the important factors in Singapore's urban development. It shapes its pleasant urban environment. The models verify that greenery, noted as GnPR, provides a good impact to the environment by reducing the air temperature. The extensive use of hard surfaces (PAVE) increases the urban air temperature by decreasing the surface moisture available for evapotranspiration. Furthermore, more solar radiation is absorbed and reradiated into heat because dry surfaces have low albedo (ALB) value (Niachou *et al.*, 2001). HBDG as a measure of building bulk is an indicator of the thermal mass of the environment which higher value shows lower temperature since it reduces the heat released to the surrounding environment. Meanwhile, large wall surface area (WALL) leads to a higher air temperature since the wall reflects short-wave and long-wave solar radiation to the environment.

Equations 1–5 above reflect the interactions between the above parameters in different time frames. T_{max}, T_{avg}, T_{min}, $T_{avg(daytime)}$ and $T_{avg(nighttime)}$ represent the peak temperature of the day (i.e. during daytime), average temperature of the day, lowest temperature of the day (i.e. during nighttime), average daytime temperature and average nighttime temperature respectively. The models (i.e. equations 1–5), when used in Geographical Information Systems (GIS), become very useful for the planners to study the air temperature conditions across the estate, either to improve the estate's existing condition or to study the air temperature impact on the future master plan. The generated temperature maps will help the planners to point out the hotspots, enabling planners to modify the estates' morphology distributions. As an example, a study was conducted to evaluate *one-north* development master plan by comparing the existing condition with the new master plan (Jusuf and Wong, 2012), summarized in Figure 3.12.

In the new master plan, *one-north* will be extensively built with a green belt, i.e. a designated green space, that splits the estate into two parts. The study simulated two models. In the first model, the green belt was simulated to be covered only with turfing with GnPR index of 1, while in the second model, the green belt would be planted with trees with GnPR index of 3. In Figure 3.13, the simulation results show the average temperature across the *one-north* estate. Comparing Figure 3.13a and b, the temperature maps show an increase of temperature especially at the northern-part of the estate, mainly due to the increase of building density and the removal of native trees in the area. The impact of green belt in the master plan model 2 seems more noticeable, creating a larger 'cool island' in the middle of *one-north* site, as shown in Figure 3.13c. Overall, the green belt in the master plan model 2 lowers the range of average air temperature to 27.7°C–28.3°C as compared with the other master plan models. The temperature is calculated at about 27.7°C in the middle of green

Figure 3.12
one-north: existing condition (a) and new master plan (b).
Maps by Jusuf.

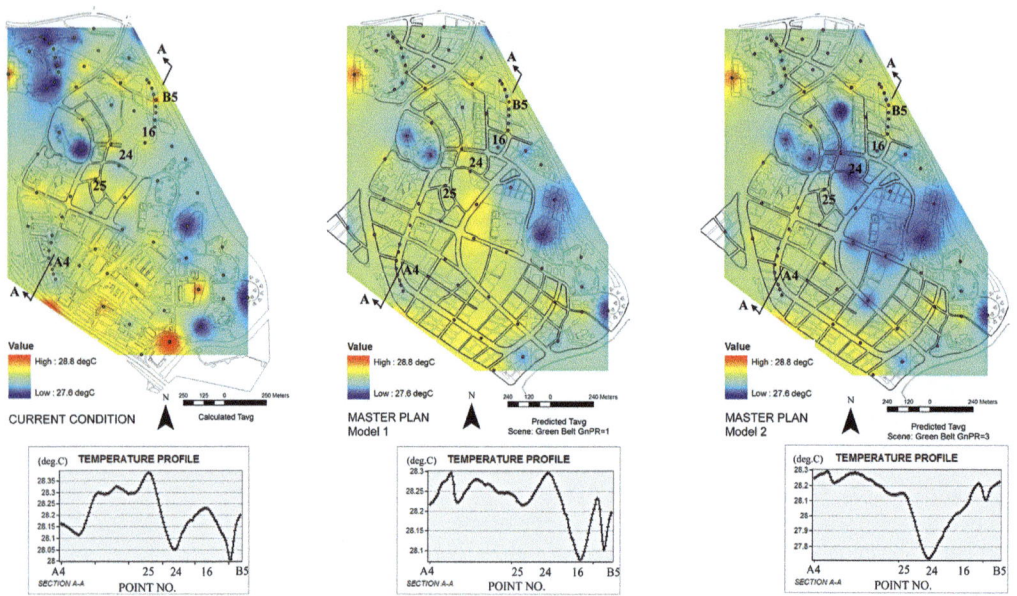

Figure 3.13 The predicted average temperature of current condition (a), master plan model 1 (b) and master plan model 2 (c).
Maps by Jusuf.

belt, i.e. Point 24. Meanwhile, the temperature map of master plan model 1 shows that grass covering green belt has no ability to provide cooling on the average air temperature.

The example above has demonstrated a simple method using the air temperature prediction models, not only to predict the temperature impact on a proposed master plan as compared with the current condition, but also to design or to set a benchmark of the green belt that is able to provide positive temperature impact to the surrounding environment. Using grass to cover the green belt has been proven to have no benefit in cooling the environment. On the other hand, planting trees and shrubs are favorable. Large urban parks can extend the positive effects to the surrounding built environment. The air temperatures have strong relationship with the density of plants, which, with higher LAIs, may cause lower ambient temperatures (Chen and Wong, 2006).

References

Akbari, H. (2002). Shade trees reduce building energy use and CO_2 emissions from power plants. *Environmental Pollution, 116*, 8119–8126.

Alexandri, E. and Jones, P. (2008). Temperature decreases in an urban canyon due to green walls and green roofs in diverse climates. *Building and Environment, 43*, 480–493.

Barrio, E. P. D. (1998). Analysis of the green roofs cooling potential in buildings. *Energy and Buildings, 27*, 179–193.

Barton, J. and Pretty, J. (2010). What is the best dose of nature and green exercise for improving mental health? A multi-study analysis. *Environmental Science and Technology, 44*(10), 3947–3955.

Bass, B. and Baskaran, B. (2003). *Evaluating rooftop and vertical gardens as an adaptation strategy for urban areas.* Institute for Research and Construction. NRCC 46737, Project number A020, CCAF Report B1046. Ottawa, Canada: National Research Council.

Bealey, W. J., McDonald, A. G., Nemitz, E., Donovan, R., Dragosits, U., Duffy, T. R. and Fowler, D. (2007). Estimating the reduction of urban PM10 concentrations by trees within an environmental information system for planners. *Journal of Environmental Management, 85*, 44–58.

Benjamin, M. T., Sudol, M., Bloch, L. and Winer, A. M. (1996). Low-emitting urban forests: A taxonomic methodology for assigning isoprene and monoterpene emission rates. *Atmospheric Environment, 30*(9), 1437–1452.

Berndtsson, J. C. (2010). Green roof performance towards management of runoff water quantity and quality: A review. *Ecological Engineering, 36*, 351–360.

Bottyan, Z. and Unger, J. (2003). A multiple linear statistical model for estimating the mean maximum urban heat island. *Theoretical and Applied Climatology, 75*, 233–243.

Ca, V. T., Asaeda, T. and Abu, E. M. (1998). Reductions in air conditioning energy caused by a nearby park. *Energy and Buildings, 29*, 83–92.

Canton, A., Cortegoso, J. L., Fernandez, J. and de Rosa, C. (2001). Environmental and energy impact of the urban forest in arid zone cities. *Architectural Science Review, 44*(1), 3–16. doi: 10.1080/00038628.2001.9697448

Chen, H., Ooka, R., Huang, H. and Tsuchiya, T. (2009). Study on mitigation measures for outdoor thermal environment on present urban blocks in Tokyo using coupled simulation. *Building and Environment, 44*, 2290–2299.

Chen, Y. and Wong, N. H. (2006). Thermal benefits of city parks. *Energy and Buildings, 38*, 105–120.

Chiesura, A. (2004). The role of urban parks for the sustainable city. *Landscape and Urban Planning, 68*, 129–138.

City of Stuttgart. (2008). *Climate booklet for urban development.* Ministry of Economy Baden-Württemberg in cooperation with Environmental Protection Department of Stuttgart. Available online at: http://www.staedtebauliche-klimafibel.de/?p=79&p2=7

Cohen, P. and Potchter, O. (2010). Daily and seasonal air quality characteristics of urban parks in the Mediterranean city of Tel Aviv. In *Proceedings CLIMAQS Workshop: Local air quality and its interactions with vegetation*, January 21–22, Antwerp, Belgium.

Conway, H. (2000). Parks and people: the social functions. In J. Woudstra and K. Fieldhouse (Eds.), *The Regeneration of Public Parks* (pp. 9–19). New York: E & FN Spon.

Corchnoy, S. B., Arey, J. and Atkinson, R. (1992). Hydrocarbon emissions from twelve urban shade trees of the Los Angeles, California, Air basin. *Atmospheric Environment*, *26B*(3), 339–348.

Corrie, C., Talbot, B., Bulkley, J. and Adriaens, P. (2005). Optimization of green roofs for air pollution mitigation. In *Proceedings of Third Annual Greening Rooftops for Sustainable Communities Conference: Awards and Trade Show*, May 4–6, Washington, DC.

D'Orazio, M., Di Perna, C. and Di Giuseppe, E. (2012). Green roof yearly performance: A case study in a highly insulated building under temperate climate. *Energy and Buildings*, *55*, 439–451.

Deutsch, B., Whitlow, H., Sullivan, M. and Savineau, A. (2005). *Re-greening Washington, DC: A green roof vision based on quantifying storm water and air quality benefits.* Retrieved from: http://www.greenroofs.org/resources/greenroofvisionfordc.pdf

Dunnett, N. and Kingsbury, N. (2008). *Planting green roofs and living walls.* Portland, USA and London, UK: Timber Press.

Fioretti, R., Palla, A., Lanza, L.G. and Principi, P. (2010). Green roof energy and water related performance in the Mediterranean climate. *Building and Environment*, *45*, 1890–1904.

Fischetti, M. (2014, June 1). The paradox of pollution-producing trees. *ScientificAmerican. com*. Retrieved September 1, 2014 from: http://www.scientificamerican.com/article/the-paradox-of-pollution-producing-trees/

Freiman, M.T., Hirshel, N. and Broday, D.M. (2006). Urban-scale variability of ambient particulate matter attributes. *Atmospheric Environment*, *40*, 5670–5684

Fujii, S., Cha, H., Kagi, N., Miyamura, H. and Kim, Y.S. (2005). Effects on air pollutant removal by plant absorption and adsorption. *Building and Environment*, *40*, 105–112.

Gaffin, S., Parshall, L., O'Keeffe, G., Braman, D. Beattie, D., and Berghage, R. (2006). Energy balance modelling applied to a comparison of white and green roof cooling efficiency. In C. Rosenzweig, S. Gaffin, and L. Parshall (Eds.), *Green roofs in the New York Metropolitan region: research report* (pp. 7–14). New York: Columbia University Center for Climate Systems Research and NASA Goddard Institute for Space Studies.

Georgi, J. N. and Dimitriou, D. (2010). The contribution of urban green spaces to the improvement of environment in cities: Case study of Chania, Greece. *Building and Environment*, *45*, 1401–1414.

Giridharan, R., Lau, S. S. Y., Ganesan, S. and Givoni, B. (2007). Urban design factors influencing heat island intensity in high rise high density environments of Hong Kong. *Building and Environment*, *42*, 3669–3684.

Gomez, F., Gaja, E. and Reig A. (1998). Vegetation and climatic changes in a city. *Ecological Engineering*, *10*, 355–360.

GRHC. (2008). *Introduction to green walls technology, benefits and design, September 2008.* Green Roofs for Healthy Cities. Retrieved October 2, 2013 from: http://www.greenscreen.com/Resources/download_it/IntroductionGreenWalls.pdf

Gromke, C. and Ruck, B. (2012). Pollutant concentrations in street canyons of different aspect ratio with avenues of trees for various wind directions. *Boundary-Layer Meteorology*, *144*(1), 41–64.

Grossi, M. (2008). Valley's greenery gives off pollutant. *Modesto Bee*. Retrieved September 1, 2014 from: http://www.modbee.com/2008/04/17/271947_valleys-greenery-gives-off-pollutant.html?rh=1

Hartig, T., Mang, M. and Evans, G. W. (1991). Restorative effects of natural environments experiences. *Environment and Behavior, 23*, 3–26.

Heisler, G. M. (1986). Effects of individual trees on the solar radiation climate of small buildings. *Urban Ecology, 9*(3–4), 337–359.

Hoyano, A. (1988). Climatological uses of plants for solar control and the effects on the thermal environment of a building. *Energy and Buildings, 11*, 181–199.

Huang, Y. J., Akbari, H., Taha, H. and Rosenfeld, H. (1987). The potential of vegetation in reducing summer cooling loads in residential buildings. *American Meteorological Society, 26*, 1103–1116.

Hutchison, B. A., Taylor, F. G., Wendt, R. L. and the Critical Review Panel. (1982). *Use of vegetation to ameliorate building microclimates: An assessment of energy conservation potentials.* Publ No. 1913. Oak Ridge National Laboratory: Environmental Sciences Division.

Jaffal, I., Ouldboukhitine, S-E. and Belarbi, R. (2012). A comprehensive study of the impact of green roofs on building energy performance. *Renewable Energy, 43*, 157–164.

Jauregui, E. (1990/91). Influence of a large urban park on temperature and convective precipitation in tropical city. *Energy and Buildings, 15–16*, 457–463.

Johnston J. and Newton J. (1993). *Building green: A guide to using plants on roofs, walls and pavements.* London, UK: London Ecology Unit. Retrieved on October 7, 2013 from: http://legacy.london.gov.uk/mayor/strategies/biodiversity/docs/Building_Green_main_text.pdf

Jusuf, S. K. and Wong, N. H. (2012). Development of empirical models for an estate level air temperature prediction in Singapore. *Journal of Heat Island Institute International, 7*(2), 111–125.

Jusuf, S. K., Wong, N. H., Hagen, E., Anggoro, R. and Hong, Y. (2007) The influence of land use on the urban heat island in Singapore. *Habitat International, 31*(2), 232–242.

Kaplan, R. (1983). The analysis of perception via preference: A strategy for studying how the environment is experienced. *Landscape and Urban Planning, 12*, 161–176.

Kawashima, S. (1990/ 1991). Effect of vegetation on surface temperature in urban and suburban areas in winter. *Energy and Buildings, 15–16*, 465–469.

Khadpekar, V.S. and Jacob, S. (2004). The urban environment. In: Kiran, B. C., Mamata, P. and Meena, R. (Eds). (2004). *Understanding environment* (pp. 175–201). New Delhi and Thousand Oaks, CA: SAGE Publications.

Kolokotroni, M., Giannitsaris, I. and Watkins, R. (2006). The effect of the London urban heat island on building summer cooling demand and night ventilation strategies. *Solar Energy, 80*(4), 383–392.

Kuttler, W. and Strassburger, A. (1999). Air quality measurements in urban green areas – A case study. *Atmospheric Environment, 33*, 4101–4108.

Lam, K. C., Ng, S. L., Hui, W. C. and Chan, P. K. (2005). Environmental quality of urban parks and open spaces in Hong Kong. *Environmental Monitoring and Assesment, 111*, 55–73.

Li, J.-F., Wai, O. W. H., Li, Y. S., Zhan, J-M., Ho Y. A., Li, J. and Lam, E. (2010). Effect of green roof on ambient CO_2 concentration. *Building and Environment, 45*, 2644–2651.

Li, W. C. and Yeung, K. K. A. (2014). A comprehensive study of green roof performance from environmental perspective. *International Journal of Sustainable Built Environment*, http://dx.doi.org/10.1016/j.ijsbe.2014.05.001.

Liesecke, H. J. (1998). Das retentionsvermogen von dachbegrunungen. *Stadt Grun, 47*, 46–53.

Lundholm, J., MacIvor, J. S., MacDougall, Z. and Ranalli, M. (2010). Plant species and functional group combinations affect green roof ecosystem functions. *PLoS ONE*, *5*(3), e9677. doi:10.1371/journal.pone.0009677.

Mahmoud, A. (2011). Analysis of the microclimatic and human comfort conditions in an urban park in hot and arid regions. *Building and Environment*, *46*(12), 2641–2656.

Makhelouf, A. (2009). The effect of green spaces on urban climate and pollution. *Iranian Journal of Environmental Health Science and Engineering*, *6*(1), 35–40.

McDonald, A. G., Bealey, W. J., Fowler, D., Dragosits, U., Skiba, U., Smith, R. I., Donovan, R. G., Brett, H. E., Hewitt, C. N. and Nemitz, E. (2007). Quantifying the effect of urban tree planting on concentrations and depositions of PM10 in two UK conurbations. *Atmospheric Environment*, *41*, 8455–8467.

McPherson, E. G., Herrington, L. P. and Heisler, M. (1988). Impacts of vegetation on residential heating and cooling. *Energy and Buildings*, *12*, 41–51.

McPherson, E. G., Simpson, J. R. and Livingston, M. (1989). Effects of three landscape treatments on residential energy and water use in Tucson, Arizona. *Energy and Buildings*, *13*, 127–138.

Meier, A. K. (1990/1991). Strategic landscaping and air-conditioning savings: literature review. *Energy and Buildings*, *15–16*, 479–486.

Mentens, J., Raes, D. and Hermy, M. (2006). Green roofs as a tool for solving the rainwater runoff problem in the urbanized 21st century. *Landscape and Urban Planning*, *77*, 217–226.

Monterusso, M. A., Rowe, D. B., Rugh, C. L. and Russell, D. K. (2004). Runoff water quantity and quality from green roof systems. *Acta Horticulturae*, *639*, 369–376.

Moran, A., Hunt, B. and Jennings, G. (2003). *A North Carolina field study to evaluate greenroof runoff quality, runoff quantity and plan growth. ASAE Paper 032303*. St. Joseph, MI: American Society of Agricultural Engineering.

Nagase, A. and Dunnett, N. (2012). Amount of water runoff from different vegetation types on extensive green roofs: Effects of plant species, diversity and plant structure. *Landscape and Urban Planning*, *104*, 356–363.

Ng, E., Chen, L., Wang, Y. and Yuan, C. (2012). A study on the cooling effects of greening in a high-density city: an experience from Hong Kong. *Building and Environment*, *47*, 256–271.

Niachou, A., Papakonstantinou, K., Santamouris, M., Tsangrassoulis, A. and Mihalakakou, G. (2001). Analysis of the green roof thermal properties and investigation of its energy performance. *Energy and Buildings*, *33*(7), 719–729.

Nowak, D. J. (2006). Institutionalizing urban forestry as a 'biotechnology' to improve environmental quality. *Urban Forestry and Urban Greening*, *5*, 93–100.

Nowak, D. J., Crane, D. E. and Stevens, J. C. (2006). Air pollution removal by urban trees and shrubs in the United States. *Urban Forestry and Urban Greening*, *4*, 115–123.

Oke, T. R. (1973). City size and the urban heat island. *Atmospheric Environment* , *7*(8), 769–779.

Oliveira, S., Andrade, H. and Vaz, T. (2011). The cooling effect of green spaces as a contribution to the mitigation of urban heat: A case study in Lisbon. *Building and Environment*, *46*, 2186–2194.

Parham, A. M. and Fariborz, H. (2010). Approaches to study urban heat island – Abilities and limitations. *Building and Environment*, *45*(10), 2192–2201.

Parizotto, S. and Lamberts, R. (2011). Investigation of green roof thermal performance in temperate climate: A case study of an experimental building in Florianópolis city, Southern Brazil. *Energy and Buildings*, *43*,1712–1722.

Parker, J. H. (1983). Landscaping to reduce the energy used in cooling buildings. *Journal of Forestry*, *81*(2), 82–84.

Perez, G., Coma, J., Martorell, I. and Cabeza, L. F. (2014). Vertical Greenery Systems (VGS) for energy saving in buildings: A review. *Renewable and Sustainable Energy Reviews*, *39*, 139–165.

Perez, G., Rincon, L., Vila, A., Gonzalez, J. M. and Cabeza, L. F. (2011). Behaviour of green façades in Mediterranean continental climate. *Energy Conversion and Management*, *52*, 1861–1867.

Perini, K., Ottele, M., Fraaij, A. L. A., Hass, E. M. and Raiteri, R. (2011). Vertical greening systems and the effects on air flow and temperature on the building envelope. *Building and Environment*, *46*, 2287–2294.

Rashid, R. and Ahmed, M. H. (2009). Thermal performance of rooftop greenery system at the tropical climate of Malaysia. *Dimensi*, *37*(1), 41–50.

Robinette, G. O. (1972). *Plant, people and environmental quality*. Washington, DC: U.S. Department of the Interior, National Park Service.

Robinson, D. and Bruse, M. (2011). Pedestrian comfort. In: D. Robinson, (Ed), *Computer modelling for sustainable urban design* (pp. 95–112). London and Washington D.C.: Earthscan.

Rosenfeld, A. H., Akbari, H., Romm, J. J. and Pomerantz, M. (1998). Cool communities: strategies for heat island mitigation and smog reduction. *Energy and Buildings*, *28*, 51–62.

Rowe, D. B. (2011). Green roofs as a means of pollution abatement. *Environmental Pollution*, *159*, 2100–2110.

Santamouris, M. (2001). The role of green spaces. In M. Santamouris (Ed.), *Energy and climate in the urban built environment* (pp. 145–159). London: James and James Science Publishers.

Santamouris, M. (2014). Cooling the cities – A review of reflective and green roof mitigation technologies to fight heat island and improve comfort in urban environments. *Solar Energy*, *103*, 682–703.

Santamouris, M., Pavlou, C., Doukas, P., Mihalakakou, G., Synnefa, A., Hatzibiros, A. and Patargias, P. (2007). Investigating and analyzing the energy and environmental performance of an experimental green roof installed in a nursery school building in Athens, Greece. *Energy*, *32*, 1781–1788.

Savio, P., Rosenzweig, C., Solecki, W. D. and Slosberg, R. B. (2006). *Mitigating New York City's heat island with urban forestry, living roofs, and light surfaces*. New York City Regional Heat Island Initiative. Albany, NY: The New York State Energy Research and Development Authority.

Schade, C. (2000). Wasserruckhaltung und Abflußbelwerte bei dunn schichtigen extensivbegrunungen. *Stadt Grun*, *49*, 95–100.

Schiller, G. and Karschon, R. (1974). Microclimate and recreational value of tree plantings in deserts. *Landscape Planning*, *1*, 329–337.

Schmidt, M. (2009). Rainwater harvesting for stormwater management and building climatization. In proceedings of *Fifth Urban Research Symposium 2009: Cities and Climate Change: Responding to an Urgent Agenda*. June 28–30, Marseille, France. Retrieved on October 7, 2013 from: http://www.gebaeudekuehlung.de/URS2009 Marseille.pdf

Schroeder, H .W. (1991). Preferences and meaning of arboretum landscapes: combining quantitative and qualitative data. *Journal of Environmental Psychology*, *11*, 231–248.

Schweitzer, O. and Erell, E. (2014). Evaluation of the energy performance and irrigation requirements of extensive green roofs in a water-scarce Mediterranean climate. *Energy and Buildings*, *68*, 25–32.

Shashua-Bar, L., Pearlmutter, D. and Erell, E. (2009). The cooling efficiency of urban landscape strategies in a hot dry climate. *Landscape and Urban Planning*, *92*, 179–186.

Shashua-Bar, L., Pearlmutter, D. and Erell, E. (2011). The influence of trees and grass on outdoor thermal comfort in a hot-arid environment. *International Journal of Climatology, 31*, 1498–1506.

Spronken-Smith, R. A. and Oke, T. R. (1998). The thermal regime of urban parks in two cities with different summer climates. *International Journal of Remote Sensing, 19*(11), 2085–2104.

Spronken-Smith, R. A., Oke, T. R. and Lowry, W. P. (2000). Advection and the surface energy balance across an irrigated urban park. *International Journal of Climatology, 20*, 1033–1047.

Streiling, S. and Matzarakis, A. (2003). Influence of single and small clusters of trees on the bioclimate of a city: a case study. *Journal Arboriculture, 29*(6), 309–316.

Sun, C-Y., Lee, K-P., Lin, T-P. and Lee, S-H. (2012). Vegetation as a material of roof and city to cool down the temperature. *Advanced Materials Research, 461*, 552–556.

Susca, T., Gaffin, S. R. and Dell'Osso, G. R. (2011). Positive effects of vegetation: urban heat island and green roofs. *Environmental Pollution, 159*, 2119–2126.

Susorova, I., Angulo, M., Bahrami, P. and Stephens, B. (2013). A model of vegetated exterior façades for evaluation of wall thermal performace. *Building and Environment, 67*, 1–13.

Taha, H. (1997). Urban climates and heat islands: albedo, evapotranspiration and anthropogenic heat. *Energy and Buildings, 25*, 99–103.

Takebayashi, H. and Moriyama, M. (2007). Surface heat budget on green roof and high reflection roof for mitigation of urban heat island. *Building and Environment, 42*, 2971–2979.

Tan, P. Y. and Sia, A. (2005). A pilot green roof research project in Singapore. In *Proceedings of Third Annual Greening Rooftops for Sustainable Communities Conference: Awards and Trade Show.* May 4–6, Washington, DC.

Thompson, C. W., Roe, J., Aspinall, P., Mitchell, R., Clow, A. and Miller, D. (2012). More green space is linked to less stress in deprived communities: evidence from salivary cortisol patterns. *Landscape and Urban Planning, 105*, 221–229.

Ulrich, R. S. (1981). Natural versus urban sciences: some psycho-physiological effects. *Environmental and Behavior, 13*, 523–556.

United Nations. (2012). *World Urbanization Prospects.* New York: Department of Economic and Social Affairs, Population Division.

VanWoert, N. D., Rowe, D. B., Andresen, J. A., Rugh, C. L., Fernandez, R. T. and Xiao. L. (2005). Green roof storm water retention: Effects of roof surface, slope and media depth. *Journal of Environmental Quality, 34*(3), 1036–1044.

Villarreal, E. L., Semadeni-Davies, A. and Bengtsson, L. (2004). Inner city storm water control using a combination of best management practices. *Ecological Engineering, 22*, 279–298.

Wong, N. H. and Chen, Y. (2009). *Tropical urban heat island.* New York: Taylor and Francis.

Wong, N. H., Chen, Y., Ong, C. L. and Sia, A. (2003). Investigation of thermal benefits of rooftop garden in the tropical environment. *Building and Environment, 38* (2), 261–270.

Wong, N. H. and Jusuf, S. K. (2010). Air temperature distribution and the influence of sky view factor in a green Singapore estate. *Journal of Urban Planning and Development, 136*(3), 261–272.

Wong, N. H. and Jusuf, S. K. (2013). *Development of climatic mapping tool for estate environmental evaluation.* Research project report. Singapore: National University of Singapore.

Wong, N. H., Jusuf, S. K., Syafii, N. I., Hajadi, N., Sathyanarayanan, H. and Manickavasagam, Y. V. (2011). Evaluation of the impact of the surrounding urban morphology on building energy consumption. *Solar Energy, 85*, 57–71.

Wong, N. H., Tan, A. Y. K, Chen, Y., Sekar, K., Tan, P. Y., Chan, D., Chiang, K. and Wong, N. C. (2010). Thermal evaluation of vertical greenery systems for building walls. *Building and Environment*, 45(3), 663–672.

Wong, N. H., Tan, A. Y. K., Tan, P. Y. and Wong, N. C. (2009). Energy simulation of vertical greenery systems. *Energy and Buildings*, 41, 1401–1408.

Yang, J., Yu, Q. and Gong, P. (2008). Quantifying air pollution removal by green roofs in Chicago. *Atmospheric Environment*, 42, 7266–7273.

Yin, S., Cai, J. P., Chen, L. P., Shen, Z. M., Zou, X. D., Wu, D. and Wang, W. (2007) Effects of vegetation status in urban green spaces on particle removal in a street canyon atmosphere. *Acta Ecologica Sinica*, 27(11), 4590–4595.

4

MITIGATING THE URBAN HEAT WITH COOL MATERIALS FOR THE BUILDINGS' FABRIC

Afroditi Synnefa[1] and Mat Santamouris[2]

[1] National and Kapodistrian University of Athens – Athens, Greece
[2] University of Athens – Athens, Greece

Introduction

Materials used in the envelope of buildings and the urban structures play a very important role in the urban thermal balance. They absorb incident solar and infrared radiation and dissipate a percentage of the accumulated heat through convective and radiative processes to the atmosphere, increasing ambient temperature. Therefore, the use of cool materials on the urban fabric that can reflect a significant part of solar radiation and dissipate the heat they have absorbed through radiation, maintaining lower temperatures compared to conventional building materials, present an effective solution to mitigate urban heat islands. Hence, the technical characteristics of the used materials determine to a high degree the energy consumption and comfort conditions of individual buildings as well as of open spaces. Among the heat island mitigation techniques that have been proposed by researchers, cool materials present some important advantages. They have an important application potential. In most urban areas, about 60 per cent of the urban fabric consists of roofs and pavements. The materials commonly used on these surfaces are characterized by low values of solar reflectance e.g. 0.2 for grey concrete and 0.05 for asphalt. They can be applied on new and existing buildings during renovation in order to avoid additional costs. They are financially viable, as their cost is comparable to conventional materials. The users do not have to change their behaviour. They are environmentally friendly, as they do not add any additional waste – on the contrary, they contribute to the reduction of waste as they prolong the lifetime of the surfaces they are applied to. (Akbari *et al.*, 1992; Akbari *et al.*, 1999; Stathopoulou *et al.*, 2009; Synnefa *et al.*, 2006).

This chapter focuses on cool materials for the building envelope with emphasis on cool roofing materials, as their impact is particularly important in mitigating urban heat. The first section describes the physical characteristics and types of cool materials for the building envelope. In the second section the basic parameters that affect the performance of cool materials are presented. Finally in the last section, the benefits and, mainly, the heat island mitigation potential of cool materials are discussed by reviewing the results of various studies from around the world.

Cool materials for the building envelope

Definitions

Cool materials are characterised by two properties:

a) High solar reflectance (SR)
b) High infrared emittance (ε)

Solar reflectance determines the ability of a surface of a material to reflect solar radiation. The term refers to the total reflectance of a surface, considering the hemispherical reflectance of radiation, integrated over the solar spectrum, including specular and diffuse reflection. It is measured on a scale of 0 to 1 (or 0–100 per cent). Infrared emittance determines the ability of a surface to release absorbed heat. It specifies how well a surface radiates energy away from itself as compared with a black body operating at the same temperature. Infrared emittance is measured on a scale from 0 to 1 (or 0–100 per cent).

Another indicator that determines how 'cool' a material is, is the Solar Reflectance Index (SRI). This is an index that combines the radiative properties (SR and ε) and convective cooling effects in one scheme. According to ASTM E1980–01 *Standard Practice for Calculating Solar Reflectance Index of Horizontal and Low-Sloped Opaque Surfaces*, SRI quantifies how hot a flat surface would get relative to a standard black (reflectivity 5 per cent, emittance 90 per cent) and a standard white surface (reflectivity 80 per cent, emittance 90 per cent). The calculation of this index is based on a set of equations (ASTM E1980–01) that requires measured values of solar reflectance and infrared emittance for specific environmental conditions. The SRI has a value of zero (for the standard black surface) and of 100 (for the standard white surface). From the definition of the SRI, it is expected that very hot materials can actually have negative values and very cool materials can have values greater than 100. Several SRI calculators have been developed and are available online (ORNL SRI Calculator, LBNL Heat Island Group SRI Calculator Excel Sheet, LEED's SRI Calculator).

A surface that is characterised by high solar reflectance and infrared emittance will remain cooler when exposed to solar radiation demonstrating lower surface temperature compared to a similar surface with lower SR and ε values. (Bretz and Akbari, 1997). Therefore, when cool materials are applied on the building envelope less heat will penetrate into the interior of the building and this cooler exterior surface will contribute to decrease the temperature of the ambient air as the heat convection intensity from a cooler surface is lower.

Types of cool materials for the building envelope

Cool materials for the building envelope consist of materials intended for the building roof and materials intended for the façades. There is extensive research related with cool roofing materials in terms of product development, applications and impact studies. On the contrary, applications and impact of cool materials for building façades are little investigated. This is mainly because of the higher mitigation potential of the cool roof surfaces as the solar radiation that arrives on vertical surfaces is lower than that on the horizontal ones. In addition, the urban geometry with the urban canyons makes the estimation of the impact of the application of solar reflective materials on vertical surfaces quite complex, and the potential problems of visual discomfort (glare) caused by highly reflective visible surfaces have limited investigation of cool façades. Several studies, however, have demonstrated a potential field interest for cool façades. Cool façades, as a cost effective technology, can contribute to achieving the environmental and energy targets related to the performance of buildings that are now required by national and EU legislation, especially for high-rise buildings, as the amount of total solar radiation on the façades is higher than that of the roof because of the building geometry. In addition the lower surface temperatures of façades covered with cool materials contribute to increasing the lifetime of the external surfaces due to less thermal stress, the performance of the insulation underneath and, in terms of urban heat mitigation, reducing the surface temperature causes a reduction of the heat released in the urban canyons at the street level.

This chapter focuses on cool roofs but also reports the findings of some studies dealing with cool façades.

Main roof products include single-ply membranes, modified bitumen roofs, coatings, built-up roofs, metal roofs, shingles and tiles. For all these roof types, there exists, or has recently been developed, a cool option. For façades, the main cool product types include coatings, plaster, etc.

Cool materials for building envelopes can be divided into two categories depending on their visible reflectance: i) White or light-colored cool materials and ii) colored cool materials. The increased interest in this technology the last few years has also led to the development of innovative cool materials with enhanced and/or combined properties: iii) innovative cool materials for the building envelope with combined properties.

i) White or light-colored cool materials

White or light-colored materials are usually applied on low-slope surfaces to avoid potential problems of glare; however, there are also cool white and light-colored products for building façades and steep-sloped roofs. Cool white coatings (usually elastomeric or cementitious) typically have solar reflectance values ranging from 0.7 to 0.85 and an emissivity about 0.85–0.90. Coating a built-up roof initially covered by a smooth black surface, with a smooth, white surface can increase the solar reflectance from 0.04 to 0.8. A grey single-ply membrane can have an initial reflectance of 0.20, while a white one can have an SR of 0.80. Modified bitumen with mineral surface has a reflectance of 0.1–0.2, but if a white coating is applied on top, the solar reflectance reaches a value of 0.65–0.7. For all these cases the emissivity is considered 0.9. White or light-color plaster that is used in vertical surfaces has a solar reflectance of 0.6–0.8 and an emissivity of 0.8–0.9. Another category of light-colored solar reflective materials is aluminum coatings, silver in color, that contain aluminum flakes in an asphalt-type resin. Aluminum flakes enhance the solar reflectance to above 0.5 for the most reflective coatings and although such value is significantly higher compared to the performance of a black material (SR = 0.05), the aluminum content has the offsetting effect of lower infrared emittance ranging usually from 0.25 to 0.65 (Synnefa et al., 2006; Berdahl and Bretz, 1997, Levinson et al., 2005a, Karlessi et al., 2014).

Table 4.1 reports typical solar reflectance, infrared emittance and SRI values of conventional and cool white materials for the building envelope.

Table 4.1 Typical solar reflectance, infrared emittance and SRI values of conventional and cool white materials for the building envelope

Conventional "hot" materials				Cool materials			
	Solar Reflectance	Infrared emittance	SRI		Solar Reflectance	Infrared emittance	SRI
Black PVC single-ply membrane	0.04–0.05	0.80–0.90	−7–0	White PVC single-ply membrane	0.65–0.85	0.80–0.90	76–107
Modified bitumen with mineral surface capsheet	0.10–0.20	0.85–0.95	4–21	Modified bitumen white coating over mineral surface	0.60–0.75	0.85–0.95	71–94
Grey concrete tile	0.18–0.25	0.85–0.90	14–25	White concrete tile	0.6–0.75	0.85–0.90	71–93
Metal roof unpainted, corrugated	0.2–0.6	0.05–0.35	−48–53	Metal roof painted with white/light colored coating	0.6–0.75	0.8–0.9	69–93
Black asphalt shingle	0.04	0.8–0.90	−7–1	White asphalt shingle	0.2–0.3	0.8–0.90	15–28
Black coating	0.04–0.05	0.8–0.9	−7–0	White coating	0.7–0.85	0.8–09	84–113

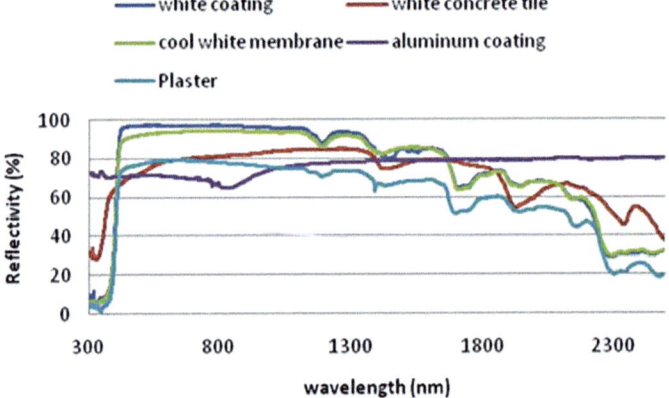

Figure 4.1
Spectral reflectance of various cool white or light colored materials

Figure 4.1 shows the spectral reflectance curves of representative cool white or light-colored materials. White materials have quite similar reflectance curves. They absorb strongly in the UV region. They have a very high reflectance in the VIS region because of the TiO_2 used and a high reflectance in the NIR part that decreases slowly with the wavelength. The absorption features in the NIR are due to vibrations of hydrogen atoms in the coatings and the C-H bond in the polymer (Berdahl and Bretz, 1997). The spectral reflectance curve of the aluminum coating is quite different. It has the tendency to increase with increasing wavelength and suddenly drop at around 800 nm.

ii) Colored cool materials

In order to avoid glare problems but also to meet aesthetic preferences, researchers have developed cool colored products that are usually applied on steep slope roofs or other visible surfaces. A cool colored surface is a surface that absorbs in the visible part of the spectrum in order to appear having a specific color and is highly reflective in the near-infrared part of the spectrum. This results in an overall higher solar reflectance compared to a colored surface of the same color (same visible reflectance) that absorbs also in the near-infrared part, taking in around about 50 per cent of the solar radiation, falls in the NIR part of the spectrum (ASTM G173–03). Several techniques have been used by researchers in order to develop cool colored materials. Synnefa et al. (2007a) have developed ten prototype cool colored coatings using near infrared reflective complex inorganic color pigments and an acryl based binder. Compared to conventionally pigmented color matched coatings, the cool colored coatings were found to have the same visible reflectance, but they higher reflectance in the near infrared part of the spectrum. The measured differences in the solar reflectance between the conventional and cool colored coatings range from 0.02 (for the light blue coating) to 0.22 (for the black coating). Levinson et al. (2007) have developed methods based on one-coat (substrate/topcoat) and two-coat (substrate/basecoat/topcoat) systems depending on the near infrared (NIR) reflectance of the substrate. For metal and clay tile products that have originally high NIR reflectance, a topcoat containing NIR reflecting pigments has been used. For gray-cement concrete tile or gray aggregate that have low NIR reflectance, a cool topcoat or basecoat with high NIR backscattering was necessary. Applying the one-coat (cool topcoat) process to metal and glazed clay-tile roofing products, they report NIR reflectances of 0.50 and 0.75, respectively. The application of a thick coating colored by rutile white NIR scattering pigments on gray-cement concrete tiles achieved NIR reflectances as high as 0.60. A two-coat process (TiO_2 rutile white basecoat + topcoat colored by NIR-transparent organic pigments) resulted in a NIR of 0.85. For asphalt shingles this method yielded a NIR of 0.45. Levinson et al. (2010a) also developed a novel technique based on a two-layer spray coating process where both layers are pigmented latex paint based on acrylic or PVDF/acrylic technology, that increases solar reflectance of concrete tiles and asphalt shingles. This factory-applied method uses as a first layer a TiO_2 rutile white and a second layer of cool color topcoat with weak NIR absorption and/or strong NIR backscattering. Forty-eight cool prototypes of concrete tiles and asphalt shingles with various shades of red, brown, green and blue have been developed with this method with solar reflectance ranging from 0.26 (dark brown) to 0.57 (light green) and for the cool colored shingles from 0.18 (dark brown) to 0.34 (light green). Ferrari et al. (2013) have investigated cool porcelain ceramic tiles trying to optimize the engobe composition and thickness for maximum reflectance. Revel et al. (2013) extended this concept also to non-white materials obtaining a series of brown tiles with increased solar reflectance. Pisello et al. (2014) have developed a new coating for clay tiles, specifically elaborated in order to optimize the infrared reflectance capability without affecting the visual appearance of the tile, using a three layer process (i) substrate, (ii) white engobe basecoat, and (iii) thin pigmented topcoat. Experimental investigation of the performance of the developed roof element showed that the elaborated coatings are able to reflect the solar radiation in the near-infrared spectrum by 75 per cent, i.e. 10 per cent more than traditional tiles, with equivalent visible appearance. The emittance did not present any substantial difference in the above cases. Revel et al. (2014) have developed of new cool colored products for the building envelope (vertical and horizontal surfaces). Black ceramic tiles for façades with improved NIR reflectance have been obtained using a double-layer approach (NIR transparent layer on highly reflecting basecoat) and both single- and double-firing processes. The double-firing process guarantees higher reflectance (NIR reflectance +0.21 and solar reflectance +0.12) thanks to the better pigment stability at relatively low temperature (1020°C) while the single-firing process (1200°C) assures an increment of reflectance (even if slightly inferior than the double firing) without increasing production costs. In parallel cool colored acrylic paints for façades and bitumen roof membranes have been developed. An improvement of NIR reflectance of +0.32 and +0.40 respectively has been measured over conventional materials of the same shade. Zinzi (2014) developed masonry paints using cool pigments. He reports that the developed paints present a significant increase of the solar and near-infrared reflectance for the dark colors with a maximum increase by a factor of 2.5 (0.16), while the increase in the case of the red coating is less significant, because the standard color already presents a higher near-infrared reflectance. The increase of the near-infrared reflectance leads to a modest emissivity reduction, with no effects for building applications. Jiang *et al.* (2014) and Qin *et al.* (2014) developed cool grey and black coatings, respectively.

iii) Innovative cool materials for the building envelope with combined properties

One example of such innovative materials with combined properties is cool and photocatalytic materials that combine high solar reflectance and infrared emittance values with photocatalytic properties, as they contain nanoparticles of TiO_2. As a result, cool and photocatalytic materials remain cooler under the sun but they also have anti-polluting, anti-soiling and disinfecting properties. Two fundamental photochemical processes that proceed on a photocatalyst's surface (e.g. TiO_2) when irradiated with ultraviolet light explain these properties. One is the photo-induced redox reaction of adsorbed substances, and the other is the photo-induced super-hydrophilicity. The combination of these two properties is also the foundation of its application in building and construction materials. (Hashimoto *et al.*, 2005). The most commonly used photocatalyst is TiO_2 (mainly the anatase type) because it presents the most efficient photo-activity, it is safe, stable, inexpensive, compatible with traditional construction materials and effective under weak solar irradiation (that contains UV radiation) in the outdoor environment (Chen and Poon, 2009). When air pollutants (NO_x, VOCs, CO etc.) arrive on the surface of a photocatalytic material, they are 'trapped' and, via redox reactions, transformed into harmless inorganic substances like carbon dioxide, nitrates, oxygen, etc., that are eliminated by the cementitious matrix of the material. This depollution has been demonstrated by many studies in the lab mainly focusing on NO_x and VOCs removal potential with promising results for NO_x and VOCs and less significant or negligible removal rates for CO. (Poon and Cheung 2006, Strini *et al.*, 2005, Berdahl and Akbari, 2008, Ibusuki 2002). Efforts have also been made through pilot projects and numerical modelling to assess the large-scale efficacy of photocatalytic materials and report significant decrease of NO_x concentration due to the use of photocatalytic materials (Yalim *et al.*, 2014; Magos *et al.*, 2008; Guerrini and Peccati 2007). When soiling (small particles and greasy deposits) occurs on the surface of a photocatalytic material, it is adhered and dissolved under the redox effect of TiO_2. Furthermore, due to super-hydrophilicity of the phenomenon of the TiO_2 surface under UV radiation, contaminants can be washed off from the surface by the rain. Several studies have demonstrated the self-cleaning properties of photocatalytic materials (Picada Project; Werle *et al.*, 2014; Aoyama *et al.*, 2014) and the disinfecting properties (Gumy *et al.*, 2006). During the last decades scientists and engineers, especially in Japan and Europe, have shown huge interest in the application of TiO_2 as photocatalyst in building materials, as indicate the large number of related patents. Today a variety of cool photocatalytic materials for building envelopes are commercially available, including coatings, tiles, paints, concrete, cement-based materials, mortar, etc. (Fujishima and Zhang 2006; Global engineering; Chen and Poon, 2009; Aïssa et al., 2011; Cheng et al., 2011, Husken and Brouwers, 2008; Diamanti et al., 2013; Sugrañez et al., 2013; Zhang et al., 2010; TX ACTIVE(r), 2006).

Another example is the development of innovative, composite cool-thermal insulating materials based on new generation of extruded polystyrene (XPS) with improved vapour permeability (lower water vapour diffusion resistance factor μ) as well as with use of a) special plasters as a final coating for vertical surfaces and b) ceramic light-colored tile for horizontal surfaces with specific features of high solar reflectance. These products aim at the combined confrontation of the two major problems that are commonly faced in the energy performance of the building envelope: the reduction of thermal losses through the opaque building elements during the winter period and the avoidance of the overheating occurring at the same elements, which has as an impact the reduction of cooling loads during the summer period (Karlessi et al., 2014).

Thermochromic materials for the building envelope are innovative materials which can change their optical and thermal properties in a dynamic way, responding to their environment, changing the color and – as a consequence, the solar properties – moving from darker to lighter tones as temperature rises. This change is reversible, which means that as temperature decreases, the material turn back to its original color. The transition is activated by the molecular structure of the pigments, which induce a spectral change of the material color. The technology relays on organic leuco-dye mixtures based on three main components which are: the color former, the color developer (usually a weak acid that causes the reversible color switch and color intensity of the final product), and the solvent. The mixture is encapsulated in microcapsules of less than microns to protect the thermochromic system from the chemicals around. Karlessi et al. (2009) have developed eleven thermochromic coatings for building envelopes using thermochromic pigments into an appropriate binder system and investigated their performance in comparison with similarly colored highly reflective (cool) and common coatings. The results demonstrated that the surface temperatures of thermochromic samples were lower than the temperatures of color-matched cool and common coatings by a maximum of 7°C and 11°C respectively. However, the application of thermochromic materials in building and urban structures presents problems in terms of durability and stability even for short outdoor exposure.

In an effort to further improve the performance of cool colored coatings, Karlessi et al. (2011) have used microencapsulated paraffin particles to develop coatings based on infrared reflective pigments doped with phase change material. Phase change materials (PCMs) are latent heat storage materials that use chemical bonds to store and release heat. Latent heat storage is preferred due to the large energy storage density and nearly isothermal nature of the storage process during which PCMs undergoes a change in phase (Zhang et al., 2007). The thermal energy transfer occurs when a material changes from solid to liquid, or liquid to solid. During daytime, the PCMs absorb part of the heat through the melting process and at night, the PCMs solidify and release the stored heat. Karlessi et al. (2011) tested 48 different coatings, including infrared reflective PCM-doped coatings, cool colored and conventional coating of the same color. All PCM-doped cool colored coatings demonstrated lower surface temperatures by a range of 3–7.5°C compared to conventional and by 1°C to 2°C compared to cool colored coating of the same color. It was also found that there is a time lag for the temperature increase of PCM-doped coatings as it stores and gives away the heat in a latent form, indicating the potential of PCM coatings when used in the building fabric to reduce and delay the peak heat load and contribute to decrease substantially surface temperatures and amortise the urban heat island phenomenon.

Finally, the use of retro-reflective materials – materials that mostly reflect the incident radiation in the same direction of the incoming radiation (i.e. the reflected beam has the same direction of the incident beam) on the building envelope – has been investigated by Rossi *et al.* (2014) in terms of optic-energy behaviour and to evaluate their potential to mitigate UHI phenomenon. It was shown that retro-reflective materials can strongly reduce mutual radiative effect among buildings located in close proximity.

Assessing the performance of cool building materials

In order to assess the performance of a cool material, it is necessary to know its solar reflectance and infrared emittance, which are the properties that determine its thermal performance. The energy balance of a surface can then be used in order to estimate the temperature that the surface will reach and, thus, have an indication of its contribution to heating the ambient air. In addition, it is necessary to have information on the degradation of the radiative properties of cool material due to its exposure in the ambient environmental condition in order to assess the long-term performance of the cool material. These issues are discussed in this section.

Measurement of the radiative properties of cool materials

In the following paragraphs the most commonly used rating techniques for cool materials for the building fabric are presented. There are several methods for measuring the solar reflectance and the infrared emittance of a surface, and this section focuses on the techniques that are adopted by the European Cool Roofs Council, the US Cool Roof Rating Council and described by specific technical standards.

Methods to measure solar reflectance

Solar reflectance, depending on the material and the specific application, can be measured using a spectrophotometer, a reflectometer and a pyranometer.

a)   The first method involves the use of a spectrophotometer equipped with an integrating sphere. This method can measure the total spectral hemispherical reflectance, as the integrating sphere collects both specular and diffuse radiation, for a small area (approximately 0.1 cm²) of a flat and uniform test sample, over the spectral range of approximately 250 to 2500 nm. Good practice procedures for the spectrophotometric measurement of the optical properties of materials are defined by ASTM E903–12, which is the standard adopted by the ECRC and the CRRC, EN14500, CIE 130–1998 and ASHRAE 74–1988. The solar reflectance can be calculated by weighted averaging, using a standard solar spectrum as the weighting function. The irradiance standard data used for this calculation are tabled in ISO 9845–1 or ASTM Standard G159–98 (replaced by ASTM G173–03). The choice of the standard solar irradiance spectrum is very important, as it can lead to differences in the determination of the solar reflectance of a sample at a range of 0–4 per cent SR and which are more important for spectrally selective materials. These differences contribute to the total uncertainty of the measurement method indicating that the use of single solar spectrum would provide comparable and 'fair' results in the framework of a product rating programme (Synnefa *et al.*, 2013).

b) Measurement of solar reflectance with a portable solar reflectometer involves the measurement of the reflectance of a flat and uniform surface of a few square centimeters. The portable solar reflectometer measures near normal–hemispherical reflectance by illuminating a surface with diffuse light and sensing light reflected at near-normal incidence. The measurement procedure is described in ASTM C1549. In Europe the use of portable reflectometer methods for measuring solar reflectance is not widespread, apart from their use in the measurement of color (Hutchins, 2009).

c) For *in situ* measurements (large surfaces) of the solar reflectance, a pyranometer can be used. The procedure is described in ASTM E1918 and requires the mounting of the pyranometer on an arm, and a stand that places the sensor at a height of 50 cm above the surface to minimize the effect of the shadow on measured reflected radiation.

Levinson *et al.* (2010b, 2010c) present a critical review of the abovementioned methods to measure solar reflectance methods and propose a clear-sky Air Mass 1 Global Horizontal spectral irradiance (AM1GH) evaluated under the atmospheric conditions specified in ASTM G173, which when used to calculate solar reflectance, better predicts solar heat gain and cool roofs energy savings.

For flat but non-uniform (heterogeneous) samples, statistical methods are needed in order to determine the solar reflectance. The CRRC-1 Test Method #1, proposed by the US Cool Roof Rating Council, uses a Portable Solar Reflectometer and requires multiple measurements at different locations on a single sample. The mean solar reflectance of the test surface is determined by averaging the solar reflectances of these randomly located spots. With this method, for samples with high degree of variation in the solar reflectance, the convergence rate is slower than typical variegated materials and requires a large sample size to estimate the solar reflectance with the required accuracy. Akbari *et al.* (2014) have proposed a Modified Monte Carlo (MMC) method that can increase the convergence rate to estimate the mean solar reflectance.

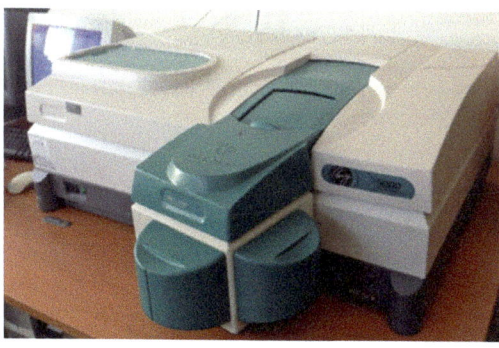

Figure 4.2
A UV-VIS-NIR spectrophotometer with an integrating
sphere (A) and a pyranometer (B)

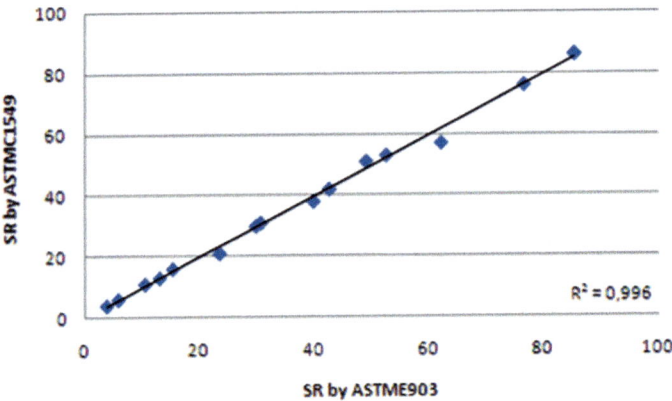

Figure 4.3 **Relation between the estimate of SR by reflectometer (ASTM C1549) and spectrophotometer (ASTM E903)**
Source: A. Synnefa, A. Pantazaras, M. Santamouris, E. Bozonnet, M. Doya, M. Zinzi, A. Muscio, A. Libbra, C. Ferrari, V. Coccia, F. Rossi, D. Kolokotsa, Interlaboratory comparison of cool roofing material measurement methods, In Proceedings of Joint Conference 34th AIVC – 3rd TightVent – 2nd Cool Roofs' – 1st Venticool, Athens, Greece, Sept. 25–26, 2013.

Additionally, for rough and/or non-uniform surfaces, the ASTM E1918 method using a pyranometer can be used and a square or round 10 m² surface is required. Akbari *et al.* (2008) have proposed a method (variant to ASTME1918) to estimate the solar reflectance of low and high-profiled tile assemblies of about 1 m² using a pyranometer and a pair of black and white masks.

Moreover, Synnefa *et al.* (2013) have conducted an interlaboratory testing aiming at investigating the suitability of different measurement methods and standards in determining the radiative properties of roofing materials. The regression analysis performed on the results showed a strong correlation between the SR determined by a spectrophotometer (ASTM E903–12) and a reflectometer (ASTM C1549). A strong correlation was also found between the determination of SR with a spectrophotometer with a large diameter integrating sphere and by both reflectometers (ASTM C1549) and spectrophotometers with a small diameter integrating sphere.

Figure 4.3 shows the correlation between the estimate of SR by reflectometer (ASTM C1549) and spectrophotometer (ASTM E903–12).

Methods to measure infrared emittance

The infrared emittance of a surface can be determined by using a portable devices (emissometers) that measure the hemispherical emittance in the range of 5–80 μm, approximately. This procedure is described in ASTM C1371–04, ASTM E408–71 and EN 1596. In addition to these methods and standards that are suitable for measuring an average infrared emittance, there are other techniques and instruments such as Fourier Transfer Infra-Red (FTIR) spectroscopy that provide detailed spectral measurement of the emittance as a function of wavelength, λ. They come mainly from the glass and glazing industry and the blinds and shutters industry (e.g. EN 12898). It should be mentioned that a lot of uncertainties are involved in the measurement of emissivity, as several factors like the sample temperature, surface geometry, etc., affect the measurement.

Synnefa *et al.* (2013) have found that the ASTM C1371–04 and EN 15976 standards give comparable results (r^2 = 0.991) for infrared emittance of flat roof products.

Assessing the thermal performance

When the radiative properties of a surface have been determined, its thermal performance can be estimated using the equation of the energy balance of the surface under the sun. For a horizontal surface (i.e. roof) this equation is the following:

$$\left(1-R\right)I = \varepsilon\sigma\left(T_s^4 - T_{sky}^4\right) + h_c\left(T_s - T_a\right) - \lambda\frac{dT}{dx}$$

where:

I:	insolation (W/m²)
σ:	the Stefan–Boltzmann constant (σ = 5,6685 x 10^{-8} W/m² K4)
h_c:	convection coefficient (W/m² K)
T_{sky}:	sky temperature (K)
T_{air}:	air temperature (K)
R:	solar reflectance or albedo of the surface
ε:	emissivity of the surface
λ:	thermal conductivity of the surface (W/mK)
dT/dx:	temperature gradient (in the x axis)

Solar radiation arrives on the surface and part of it is reflected and part absorbed by the surface contributing to its heating. Consequently, the surface emits radiation in the far-infrared part of the spectrum as radiation exchange occurs between two surfaces (in this case, roof and 'sky') when one is warmer than the other and they 'view' each other. In addition, the roof surface exchanges energy by convection with the air above the roof. Finally, heat is conducted through the layers within the roof (insulation, etc.) from the warmer side to the cooler side.

In order to assess the thermal performance of a surface on which a cool material has been applied, surface temperature measurements should be conducted. An appropriate surface temperature sensor should be selected and connected to a data logging system to have continuous measurements for detailed analysis. Attention should be paid when placing the sensor on the surface, as the sensor should not be exposed directly to incident solar radiation and the optical characteristics of the sensor upper surface should be similar to those of the surface tested. In addition an infrared thermometer can be used to take spot measurements of the surface temperature and an average value can be calculated in order to determine the temperature of a large surface (e.g. a roof). An infrared camera can be also be used in order to observe the temperature distribution on the surface of the samples as well as to depict the temperature differences between the samples.

A large number of experimental studies have been conducted in order to evaluate the thermal performance of conventional and cool materials for the building fabric. These studies have been conducted mainly in Europe and the US under

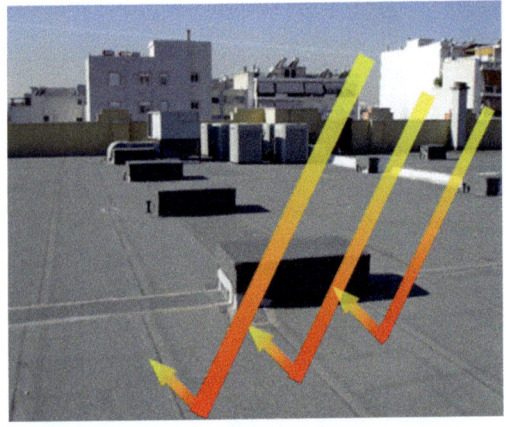 

 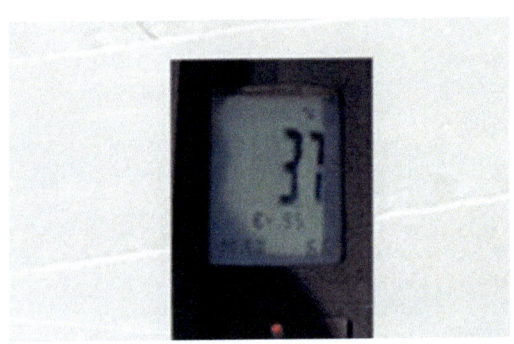

Figure 4.4
The surface temperature of a roof covered with a cool white coating (B) is significantly lower than the surface temperature of a roof covered with a conventional dark color asphalt membrane

summer conditions and the results concerning surface temperatures can be grouped into three categories:

a) Surfaces with low solar reflectance and high infrared emittance (e.g. black coating, asphalt shingle, black gravel surface) have SRI values ranging from –7–20 and can reach temperatures as high as 75–80°C,

b) surfaces with medium to high solar reflectance and low infrared emittance (e.g. unpainted metal roofs, aluminum coatings) have SRI values ranging from –48–72 and can reach temperatures as high as 60–75°C and

c) surfaces with high solar reflectance and infrared emittance (cool white coatings, white membranes etc.) have SRI values ranging from 70–113 can reach temperatures of averagely 45°C, depending of course on local ambient conditions.

The temperature differences between cool colored and conventional materials of the same color depend as expected on the difference in their solar reflectance values. Various studies report that for summer conditions, a reduction in the temperature of the surface covered with the cool colored material ranging from 5–14°C and corresponding to an increase in the solar reflectance of 0.15–0.47 (Synnefa et al., 2007a; Levinson et al., 2007; Miller et al., 2005). During winter, this temperature difference between cool and standard colored coatings diminishes significantly because of the lower amount of solar radiation arriving on the surfaces, which is desirable in order to avoid any heating penalty (Synnefa et al., 2007a).

Ageing of cool materials for the building envelope

In order to assess the long-term performance of cool materials for the building envelope, it is important to know the aged values of solar reflectance and infrared emittance, as these radiative properties may change over time due to their exposure to ambient conditions. The reduction of the solar reflectance values due to ageing can result in a significant reduction of the cool roof cooling load savings (Mastrapostoli *et al.*, 2014; Paolini *et al.*, 2014). Ageing of materials is caused by two processes: weathering and soiling. Weathering includes exposure to solar radiation (mainly ultraviolet [UV] radiation), moisture, heat and temperature fluctuations, and wind. Soiling processes include deposition of atmospheric pollutants (e.g. black carbon, dust, and organic and inorganic particulate matter), as well as microbiological contamination (Berdahl *et al.*, 2008).

As a general rule of thumb, exposure tends to moderately decrease the reflectance of light-colored materials, while moderately increasing the reflectance of dark materials. LBNL suggests a simple expression to estimate the aged solar reflectance of a roof:

$$SR_{aged} = \rho_o + c\left(SR_{initial} - \rho_o\right)$$

where constants $\rho_o = 0.2$ and $c = 0.7$. That is, the change to reflectance with ageing is modeled as a 30 per cent reduction in the difference between the initial reflectance and a value of 0.2. (Levinson *et al.*, 2005a). However, in order to have an accurate and fair comparison of the ageing performance between different cool roof products (as different products of even the same material type tend to have different ageing performances), it is necessary to adopt specific experimental procedures for this evaluation.

The most commonly used experimental procedures for determining the ageing of roofing materials properties consist of the following exposure methods (Jacques, 2000, Johnson and McIntyre 1996):

a) **Natural weathering**: It consists of exposing samples to outdoor ambient conditions (direct sunlight and other elements of weather). There are a number of exposure sites (weathering farms) around the world, each with different climatic conditions, which can be used for the purpose of predicting the exterior durability of materials. Elements like angle of exposure affect significantly the results and should be clearly defined. It is important, during natural exposure evaluations, to monitor all related environmental conditions, e.g. ambient weather conditions, radiation exposure, pollutant concentrations, etc.

b) **Artificial weathering**: This type of weathering tests is used in order to accelerate the degradation and study the material's behaviour under controlled environmental conditions in the lab and in a reasonably fast time. There is a large number of commercial artificial weathering acceleration procedures available involving the use of light sources, like the xenon arc spectrum modified with specific filters, which gives a good simulation of sunlight, fluorescent lamps, etc.

There is also the method of accelerated weathering involving outdoor exposure of materials accelerated by the use of high levels of solar irradiance concentrated by a Fresnel array of highly efficient reflectors onto the surface of the sample.

However this method is not commonly used for the assessment of ageing of cool materials. Moreover, in order to meet the industry's demand for a method that provides accurate ageing results in a shorter time, which would expedite the introduction of new and better performing products to the market and help code bodies implement cool roof requirements, Sleiman *et al.* (2014) have developed a novel laboratory accelerated ageing method that combines soiling and weathering processes and simulates 3 years of outdoor exposure in a few days. This accelerated ageing method is under consideration as an ASTM standard.

There are several standards related to the weathering of materials (e.g. ASTM G7–97, ASTM D1669, ISO 11341). More specifically, for cool materials, the US CRRC in the product rating program CRRC-1 requires the determination of the aged radiative properties of roofing materials to be reported. Aged testing is a mandatory requirement and all materials should be sent to approved test farm facilities where they shall undergo weathering exposure for three years according to the provisions defined. Ageing testing is also mandatory in order for a cool roofing material to receive Energy Star labelling. Three years exposure is required for the tested products either a) on three existing roofs where the material is installed, one of them being located in a large metropolitan area. Measurements for the determination of the aged properties can be done on site or on samples taken from the roof according to specific guidelines; b) by outdoor exposure on accredited commercial or private weathering farms; or c) by using the CRRC Color Family Program as defined in the Product Rating Program Manual (CRRC-1).

Several studies report the ageing of cool roofing materials over time. Sleiman *et al.* (2011) have evaluated solar reflectance losses after three years of natural exposure reported in two separate databases: the Rated Products Directory of the US Cool Roof Rating Council (CRRC) and information reported by manufacturers to the US Environmental Protection Agency (EPA) ENERGY STAR® rating program. They found that products with high initial solar reflectance tended to lose reflectance, while those with very low initial solar reflectance tended to become more reflective as they aged. Also, it was observed that absolute solar reflectance losses for samples of medium-to-high initial solar reflectance were 2–3 times greater when exposed to Florida (hot and humid) than in Arizona (hot and dry); losses in Ohio (temperate but polluted) were intermediate. They also disaggregated results by product type: factory-applied coating, field-applied coating, metal, modified bitumen, shingle, single-ply membrane and tile. They concluded that that absolute solar reflectance losses were largest for field-applied coating, modified bitumen and single-ply membrane products, and smallest for factory-applied coating and metal products. Finally, they found that the 2008 Title 24 provisional aged solar reflectance formula over-predicts the measured aged solar reflectance of 0–30 per cent of each product type. The rate of overprediction was greatest for field-applied coating and single-ply membrane products and least for factory-applied coating, shingle, and metal products. They propose the use of product-specific formulas in the form

$$SR_{aged} = 0.2 + b\left(SR_{in} - 0.2\right)$$

in order to have a provisional aged solar reflectance estimation until values after exposure are available.

Research has also shown that loss in the solar reflectance of samples exposed to outdoor conditions is more significant during the first six months of exposure and that the emissivity is not significantly affected by ageing. (Bretz and Akbari, 1997; Synnefa et al., 2006). Cleaning the surface of aged cool roof products can restore its initial solar reflectance (up to a percentage) depending on the product type, the ageing mechanisms that have caused the loss of reflectance and the cleaning method used (Berdahl et al., 2002; Levinson et al., 2005b; Santamouris et al., 2008).

Quantifying the mitigation potential of cool roofs

This chapter describes the benefits from the use of cool materials for the building envelope for the building users and the environment, at building, city and global scales, and quantifies their impact in mitigating the problem of overheating of urban areas.

Benefits of cool materials for the building envelope at building scale

The impact of cool materials at building scale can be quantified both experimentally and via numerical modelling. The experimental assessment consists of the following steps after the identification of the building on which the cool materials will be applied:

a) Building audit and data collection (weather conditions, air temperature inside the building, surface temperatures on and below the application surface (e.g. roof), AC and total building power consumption, operation schedules and other building configuration and use data, etc.), at least for one full year including pre- and post-intervention periods.

b) Measurement of solar reflectance and emissivity of the application surface before and after the intervention.

c) Data analysis and development of statistical model to relate AC energy use and demand to outdoor-indoor air temperatures difference, calculation of AC daily energy savings and demand reduction.

d) Cost benefit analysis: Annual energy and demand savings (euros/m²), payback period, future savings in order to assess the benefits of cool materials at building scale.

In order to assess the impact of the application of cool materials on the building envelope by using building simulation models the following process is usually followed: a) Collection of all available data on the building (plans, building configuration, building use, systems, bills, etc.); b) data on initial and final roof solar reflectance and emissivity; c) measured data, if possible, to calibrate and validate the model; d) data analysis and calculation of energy savings. Possibility to test more building variants (increased insulation, other system efficiencies, etc.), compare with other energy saving measures; e) Cost-benefit analysis: annual energy and demand savings (euros/m²), payback period, future savings. This methodology uses dynamic simulations software (e.g. TRNSYS, Energy Plus, etc.) that give detailed results and are time consuming. There are several online calculators allowing the estimation of annual energy and monetary savings associated with choosing a cool roof instead of a dark roof (e.g. Roof Savings Calculator, DOE Cool Roof Calculator, EU Cool Roofs Toolkit).

A number of experimental and theoretical studies report the impact of cool building envelopes on thermal comfort conditions, energy consumption and lifetime of the cool surface. The following paragraphs summarise these findings.

Improvement of thermal comfort conditions inside the building: If a building is not air-conditioned, the reduced heat transfer from the cooler surface (e.g. roof) results in lower indoor temperatures and improved thermal comfort conditions. Various monitoring and simulation campaigns indicate decrease of the indoor temperature ranging averagely from 1–3°C and significant reduction of the discomfort hours (Gartland 1998; Synnefa *et al.*, 2007b; Synnefa *et al.*, 2012; Pisello *et al.*, 2014).

Reduction of cooling energy use: A large number of experimental and simulation studies have been performed in residential and non-residential buildings documenting savings that vary averagely from 10 per cent to 40 per cent depending on a lot of factors like local climatic conditions, building configuration and use, etc. Air conditioning savings are more important for: hot climatic conditions, low levels of insulation, large roof surface compared to other surfaces of the building (Haberl and Cho, 2004; Levinson *et al.*, 2005a; Akbari *et al.*, 2005; Synnefa *et al.*, 2007b; Synnefa *et al.*, 2012; Romeo and Zinzi, 2013). Studies have shown that this increase is far less important than the corresponding cooling energy savings, resulting into positive net savings for warm/moderate climatic conditions. This is explained by the fact that during winter, the sun is much lower in the sky and solar radiation arrives to a horizontal surface less intensely. There is a higher probability of overcast skies and there is less solar availability (fewer hours of sunshine), so less total energy arrives on a surface to be absorbed or reflected over the same period of time as during the summer (Synnefa *et al.*, 2007b; Synnefa *et al.*, 2012; Akbari and Hosseini, 2014).

Reduction of peak energy demand for cooling and associated costs: The peak cooling energy savings from cool roofs range between 3 per cent and 35 per cent. Lower peak demand not only saves on total electrical use but also can reduce demand fees that some utilities charge commercial and industrial building owners (Akbari *et al.*, 1997; Levinson *et al.*, 2005a).

Increased lifetime of the cool surface: Many studies have shown that the application of a cool material replacing a conventional 'hot option' results in lower surface temperatures as high as 30°C and significantly reduce the diurnal temperature fluctuations (Synnefa *et al.*, 2012; Bozonnet *et al.*, 2011; Romeo and Zinzi, 2013). The degradation of materials is associated with chemical reactions that progress faster with higher temperatures. Furthermore, tempera-ture swings impose stresses due to differential thermal expansion. Therefore, the roof surface suffers from less thermal fatigue, suggesting a longer lifetime of the cool surface (Levinson *et al.*, 2005a, Berdahl *et al.*, 2008).

Benefits of cool materials for the building envelope at city scale

Mitigation of the UHI

In order to estimate the impact from the city-wide application of cool materials as well as other surface modification scenarios (e.g. increase of vegetation cover) researchers have used mesoscale (and fine resolution meso-urban) meteoro-logical modelling. Large-scale increases of urban albedo can affect the urban microclimate by lowering air temperatures, due to less heat transfer from a cooler surface to the ambient air, and thus, mitigating the heat island effect. In most such studies the increase of the city's albedo leads to a reduction of the air temperature by averagely 1–3°C (Sailor, 1995; Taha, 1997, 2005, 2008a, 2008b; Taha *et al.*, 1999, 2000, 2002; Synnefa *et al.*, 2008; Marchesi *et al.*, 2014).

Observation data and analysis at local scale provided by Campra *et al.* (2008) conclude that widespread adoption of white-roofed greenhouse farming has resulted in local climate cooling in southeastern Spain.

Reduction of cooling energy use and peak cooling loads

In order to investigate the performance of high albedo surfaces in energy savings of buildings in urban areas, most studies have used meso- and microscale models coupled to BEMs. The decrease in air temperature resulting from the large-scale use of cool materials can lead to a reduction of cooling energy use and peak cooling loads (Akbari and Taha, 1992; Rosenfeld *et al.*, 1998; Taha *et al.*, 1999, 2000; Akbari and Konopacki 2004). Akbari and Konopacki (2005) have calculated the cooling energy savings due to the application of heat island mitigation strategies (application of cool materials and increase in vegetation cover) for 240 regions in the United States. It was found that for residential buildings, the cooling energy savings vary between 12 per cent and 25 per cent, for office buildings between 5 per cent and 18 per cent, and for commercial (retail stores) buildings between 7 per cent and 17 per cent. Akbari *et al.* (2014) have also investigated the effect of increasing the albedo of urban surfaces in a cold climate region (Montreal, Canada) by coupling mesoscale, urban canopy, and building energy models. Increasing urban albedo showed little effects on net annual cooling and heating energy use (consistent with other studies). They concluded that the net conditioning energy use increased by 1 per cent. However, they found that increasing the albedo in Montreal decreased the air temperature during summer time by as much as 3°C and this temperature reduction can have significant effect on urban air quality and summer heat waves.

Reduced air pollution

Air temperature decrease from UHI mitigation can potentially lead to reduced air pollution and greenhouse gases emissions both directly and indirectly (Rosenfeld *et al.*, 1995, 1998). Direct reduction (Rosenfeld *et al.*, 1995, 1998) of air pollution is due to the fact that less cooling energy is used; therefore fewer power plant emissions are produced (CO_2, NO_x, and PM10 particles). Indirect air pollution reductions reflect the fact that the reaction of ozone formation (that produces smog) accelerates at higher temperatures, therefore at lower urban air temperatures the probability of smog formation is decreased (Taha, 1997). More specifically, a number of measurement and simulation studies have demonstrated that decreasing the air temperature of Los Angeles by 1.5–2°C using heat island mitigation strategies results in a reduction of 10–20 per cent in population weighted smog (ozone) (Taha *et al.*, 1994, 1999, 2000). Simulations with state-of-science meteorological, emissions, and photochemical models report positive (beneficial) and negative (inadvertent) impacts of urban heat island mitigation on ozone formation (Taha, 2005, 2007, 2008a). It should be pointed out, however, that the occurrence of elevated ozone concentrations is a regional issue and although there is a threshold temperature above which the likelihood of smog events increases, other factors such as atmospheric and surface transport mechanisms greatly influence the ozone distribution (Gray and Finster, 1999).

Benefits of cool materials for the building envelope at the global scale

At the global scale, Akbari *et al.* (2009) have calculated that widespread adoption of high albedo structural surfaces ('cool roofs' and 'cool pavements' that would increase albedo of urban areas 0.1) in low-and mid-latitude cities worldwide would generate a significant negative radiative forcing at a global scale, and they estimate that this could potentially contribute to mitigating global warming effect by offsetting the equivalent of 44 Gt of CO_2 emissions. Menon *et al.* (2010) have quantified the change in radiative forcing and land surface temperature that may be obtained by increasing the albedos of roofs and pavements in urban areas in temperate and tropical regions of the globe by 0.1, using a land surface model coupled to the GEOS-5 Atmospheric General Circulation Model. They found that the global average increase in the total outgoing radiation was 0.5 W/m^2, and temperature decreased by approximately 0.008K for an average 0.003 increase in surface albedo. Based on these forcings, the expected emitted CO_2 offset for a plausible 0.25 and 0.15 increase in albedo of roofs and pavements, respectively, for all global urban areas, was found to be approximately 57 Gt of CO_2. Oleson *et al.* (2010) have quantified the effects of white roofs on urban temperature, using a global climate model coupled with an urban canyon model and found a decrease in urban daily maximum temperature by 0.6 K and daily minimum temperature by 0.3 K. Cotana *et al.* (2014) showed that the introduction of albedo control technologies as a method for global warming mitigation and the assignment of a tradeable economic value (e.g. emissions credits) could represent a key strategy for territories located in tropical and mid-latitude areas, especially those characterized by low cloudiness.

Conclusions

Cool materials are a cost-effective, environmentally friendly and passive technique that contributes to achieving energy efficiency in buildings by lowering energy demand for cooling and improving the urban microclimate by lowering surface and air temperatures. This chapter has summarised the basic aspects related to cool materials for the urban fabric. The radiative properties that define cool materials have been described and the different types of materials that are commercially available or under development as a cool solution for the building envelope have been reported. In addition, the parameters affecting the performance of cool materials have been analysed. The commonly used methods to measure the radiative properties of cool materials have been described and information was given on how to assess the thermal performance of such materials theoretically and experimentally. The impact of ageing of cool materials has also been analysed. The benefits arising from the use of cool materials on the building envelope at building, city and global scale have been described by reporting relevant research results.

References

Aïssa A.H., Puzenat E., Plassais A., Herrmann J.-M., Haehnel C., Guillard C., 2011. Characterization and photocatalytic performance in air of cementitious materials containing TiO_2. Case study of formaldehyde removal. *Applied Catalysis B: Environmental, 107*: 1–8.

Akbari H., Bretz S., Kurn D., Hanford J., 1997. Peak power and cooling energy savings of high-albedo roofs. *Energy and Buildings 25*: 117–126.

Akbari H., Davis S., Dorsano S., Huang J., Winert S., 1992. *Cooling our communities – A guidebook on tree planting and white coloured surfacing*. Washington, DC: US Environmental Protection Agency, Office of Policy Analysis, Climate Change Division.

Akbari H., Hooshangi H.R., Touchaei A.G., 2014. Measuring solar reflectance of variegated flat roofing materials using modified Monte Carlo method. In *Proceedings of the Third International Conference on Countermeasures to Urban Heat Island*. Venice, Italy. October 13–15.

Akbari H. and Hosseini M., 2014. Cool roofs in cold climates: Savings or penalties?. In *Proceedings of the Third International Conference on Countermeasures to Urban Heat Island*. Venice, Italy. October 13–15.

Akbari H. and Konopacki S., 2004. Energy effects of heat-island reduction strategies in Toronto, Canada. *Energy 29*: 191–210.

Akbari H. and Konopacki S., 2005. Calculating energy-saving potentials of heat-island reduction strategies. *Energy Policy 33*: 721–756.

Akbari H., Levinson R., Miller W., Berdahl P., 2005. Cool colored roofs to save energy and improve air quality. *In: Proc. International Conference: Passive and Low Energy Cooling for the Built Environment*. Santorini, Greece. May.

Akbari H., Levinson R., Stern S., 2008. Procedure for measuring the solar reflectance of flat or curved roofing assemblies. *Solar Energy 82*(7): 648–655.

Akbari H., Menon S., Rosenfeld A., 2009. Global cooling: Increasing world-wide urban albedos to offset CO_2. *Climatic Change 94*: 275–286.

Akbari H., Rose L.S., Taha H., 1999. *Characterizing the fabric of the urban environment: A case study of Sacramento, California*. LBNL-44688. Berkeley, California: Lawrence Berkeley National Laboratory.

Akbari H. and Taha H., 1992. "The Impact of Trees and White Surfaces on Residential Heating and Cooling Energy Use in Four Canadian Cities", *Energy, The International Journal 17*(2): 141–149.

Aoyama T., Sonoda T., Hamamura T., Nakanish Y., Peeters L., Takebayashi H., 2014. Development of self-cleaning top-coat for cool roof. In *Proceedings of the Third International Conference on Countermeasures to Urban Heat Island*. Venice, Italy. October 13–15.

ASHRAE 74–1988, 1988. *Method of measuring solar optical properties of materials*. Atlanta, GA: ASHRAE.

ASTM C1371–04, 2004. *Standard test method for determination of emittance of materials near room temperature using portable emissometers*. West Conshohocken, PA: ASTM International.

ASTM C1549, 2002. *Standard test method for determination of solar reflectance near ambient temperature using a portable solar reflectometer*. West Conshohocken, PA: ASTM International.

ASTM D1669, 2003. *Standard practice for preparation of test panels for accelerated and outdoor weathering of bituminous coatings*. West Conshohocken, PA: ASTM International.

ASTM E1918, 1997. *Standard test method for measuring solar reflectance of horizontal and low-sloped surfaces in the field*. West Conshohocken, PA: ASTM International.

ASTM E1980–01, 2001. *Standard practice for calculating solar reflectance index of horizontal and low-sloped opaque surfaces*. West Conshohocken, PA: ASTM International.

ASTM E408–71, 2008. *Standard test methods for total normal emittance of surfaces using inspection-meter techniques*. West Conshohocken, PA: ASTM International.

ASTM E903–12, 2012. *Standard test method for solar absorptance, reflectance, and transmittance of materials using integrating spheres*. West Conshohocken, PA: ASTM International.

ASTM G173–03, 2008. *Standard tables for reference solar spectral irradiances: Direct normal and hemispherical on 37 tilted surface*. West Conshohocken, PA: ASTM International.

ASTM G7–97, 1997. *Standard practice for atmospheric environmental exposure testing of nonmetallic materials*. West Conshohocken, PA: ASTM International.

ASTM Standard G159–98, 1998. *Standard tables for references solar spectral irradiance at air mass 1.5: Direct normal and hemispherical for a 37 tilted surface*. West Conshohocken, PA: ASTM International.

Berdahl P. and Akbari H., 2008. Evaluation of titanium dioxide as a photocatalyst for removing air pollutants CEC-500-2007-112. Berkeley, CA: California Energy Commission, PIER Energy-Related Environmental Research Program.

Berdahl P., Akbari H., Levinson R., Miller W.A, 2008. Weathering of roofing materials – an overview. *Construction and Building Materials 22*: 423–433.

Berdahl P., Akbari H., Rose L.S., 2002. Aging of reflective roofs: soot deposition. *Applied Optics 41*: 2355–2360.

Berdahl P. and Bretz S.E., 1997. Preliminary survey of the solar reflectance of cool roofing materials. *Energy and Buildings 25*: 149–158.

Bozonnet E., Doya M., Allard F., 2011. Cool roofs impact on building thermal response: A French case study. *Energy and Buildings 43*(11): 3006–3012.

Bretz S., Akbari H., 1997. Long-term performance of high albedo roof coatings. *Energy and Buildings 25*: 159–167.

Campra P., Garcia M., Canton Y., Palacios-Orueta A., 2008. Surface temperature cooling trends and negative radiative forcing due to land use change toward greenhouse farming in southeastern Spain. *J. Geophys. Res. 113*: D18109. doi:10.1029/2008 JD009912.

Chen J. and Poon C., 2009. Photocatalytic construction and building materials: From fundamentals to applications. *Building and Environment 44*(9): 899–1906

Cheng M.-D., Pfiffner S.M., Miller W.A., Berdahl P., 2011. Chemical and microbial effects of atmospheric particles on the performance of steep-slope roofing materials. *Building and Environment 46*: 999–1010. doi:10.1016/j.buildenv.2010.10.025

CIE 130–1998, 1998. *Practical methods for the measurement of reflectance and transmittance*. Vienna, Austria: International Commission on Illumination.

Cotana F., Rossi F., Filipponi M., Coccia V., Pisello A. L., Bonamente E., Petrozzi A., Cavalaglio G., 2014. Albedo control as an effective strategy to tackle global warming: A case study. *Applied Energy 130*: 641–647.

CRRC (USA): Product Rating Program CRRC-1, 2008. Cool Roof Rating Council Inc. Available online: http://www.coolroofs.org/documents/CRRC-1_Program_Manual_090810.pdf. Accessed December 2014.

Diamanti M.V., Del Curto B., Ormellese M., Pedeferri M.P., 2013. Photocatalytic and self cleaning activity of colored mortars containing TiO$_2$. *Construction and Building Materials 46*: 167–174.

DOE Cool Roof Calculator. Available online: http://www.ornl.gov/sci/roofs+walls/facts/CoolCalcEnergy.htm. Accessed December 2014.

EN 12898, 1998. *Glass in building – Determination of the emissivity*. London: BSI.

EN 14500, 2008. *Blinds and shutters – Thermal and visual comfort – Test and calculation methods*. London: BSI.

EN 15976, 2011. *Flexible sheets for waterproofing. Determination of emissivity*. London: BSI.

Energy Star. Energy Star program requirements for roof products. Available online: http://www.energystar.gov/ia/partners/product_specs/program_reqs/roofs_prog_req.pdf. Accessed December 2014.

EPA, 2009. Reducing Urban Heat Islands: Compendium of Strategies Cool Roofs. Available online: http://www.epa.gov/hiri/resources/compendium.htm. Accessed December 2014.

EU Cool Roofs Toolkit. Available online: http://pouliezos.dpem.tuc.gr/coolroof/cool calcenergy_eu.html. Accessed December 2014.

Ferrari C., Libbra A., Muscio A., Siligardi C., 2013. Design of ceramic tiles with high solar reflectance through the development of a functional engobe. *Ceramics International* 39: 9583–9590

Fujishima A. and Zhang X.T., 2006. Titanium dioxide photocatalysis: Present situation and future approaches. *Comptes Rendus Chimie* 9(5–6): 750–760

Gartland L., 1998. Roof coating evaluation for the Home Base Store in Vacaville, California. Oakland, CA: PositivEnergy.

Global engineering. Available online: http://www.globalengineering.info/index.php?idx= prodotti. Accessed December 2014.

Gray K.A. and Finster M.E., 1999. The urban heat island, photochemical smog, and Chicago: Local features of the problem and solution. Northwestern University Tech. Rep., Atmospheric Pollution Prevention Division. Evanston, IL: Northwestern University. Available online: http://www2.epa.gov/heat-islands/urban-heat-island-photochemical-smog-and-chicago-local-features-problem-and-solution. Accessed December 2014.

Guerrini G.L. and Peccati E., 2007. Photocatalytic cementitious roads for depollution. In P. Baglioni and L. Cassar, (eds.) *RILEM International Symposium on Photocatalysis, Environment and Construction Materials* (pp. 179–186). Italy: RILEM.

Gumy D., Morais C., Bowen P., Pulgarin C., Giraldo S., Hajdu R., Kiwi J., 2006. Catalytic activity of commercial of TiO_2 powders for the abatement of the bacteria (*E. coli*) under solar simulated light: Influence of the isoelectric point. *Applied Catalysis B: Environmental* 63: 76–84.

Haberl J. and Cho S., 2004. *Literature review of uncertainty of analysis methods (cool roofs), Report to the Texas Commission on Environmental Quality*. College Station, TX: Energy Systems Laboratory, Texas A&M University.

Hashimoto K., Irie H., Fujishima A., 2005. TiO_2 photocatalysis: A historical overview and future prospects. *Japanese Journal of Applied Physics* 44: 8269–8285

Husken G. and Brouwers H J.H., 2008. Air purification by cementitious materials: Evaluation of air purifying properties. *In Proceedings of the International Conference on Construction and Buildings Technology* (pp. 263–274). Malaysia: ICCBT.

Hutchins M., 2009. *Progress report of the EU Cool Roofs Council Technical Committee*. Available online: http://www.coolroofs-eu.eu/. Accessed December 2014.

Ibusuki T., 2002. Cleaning atmospheric environment. In M. Kaneko and I. Okura (eds.), *Photocatalysis: Science and technology* (pp. 123–159). Tokyo: Kodansha Ltd.

ISO 11341: *Paints and varnishes – Artificial weathering and exposure to artificial radiation Exposure to filtered xenon arc radiation*. Switzerland: International Organization for Standardization.

Jacques L., 2000. Accelerated and outdoor/natural exposure testing of coatings. *Progress in Polymer Science* 25: 1337–1362.

Jiang L., Xue X., Qu J., Qin J., Song J., Shi Y., Zhang W., Song Z., Li J., Guo H., Zhang T., 2014. The methods for creating energy efficient cool gray building coatings – Part II: Preparation from pigments of complementary colors and titanium dioxide rutile. *Solar Energy Materials and Solar Cells* 130: 410–419.

Johnson B.W. and McIntyre R., 1996. Analysis of test methods for UV durability predictions of polymer coatings. *Progress in Organic Coatings* 27: 95–106.

Karlessi T., Santamouris M., Apostolakis K., Synnefa A., Livada I., 2009. Development and testing of thermochromic coatings for buildings and urban structures. *Solar Energy* 83(4): 538–551.

Karlessi T., Santamouris M., Chadiarakou S., Kontos M., Leonidaki K., Antoniadou P., Boemi S. N., Anastaselos D., Papadopoulos A., Alexopoulos E., 2014. Towards the assessment of planning, developing and producing of innovative composite cool-thermal insulating materials. In *Proceedings of the Third International Conference on Countermeasures to Urban Heat Island*. Venice, Italy. October 13–15.

Karlessi T., Santamouris M., Synnefa A., Assimakopoulos D., Didaskalopoulos P., Apostolakis K., 2011. Development and testing of PCM doped cool colored coatings to mitigate urban heat island and cool buildings. *Building and Environment 46*(3): 570–576.

LBNL Heat Island Group SRI Calculator Excel Sheet. Available online: http://coolcolors.lbl.gov/assets/docs/SRI per cent20Calculator/SRI-calc10.xls. Accessed December 2014.

LEED's SRI Calculator. Available online: http://www.usgbc.org/DisplayPage.aspx?CMSPageID=1447. Accessed December 2014.

Levinson R., Akbari H., Berdahl P., Wood K., Skilton W., Petersheim J., 2010a. A novel technique for the production of cool colored concrete tile and asphalt shingle roofing products. *Solar Energy Materials and Solar Cells 94*(6): 946–954.

Levinson R., Akbari H., Berdahl P., 2010b. Measuring solar reflectance – Part I: Defining a metric that accurately predicts solar heat gain. *Solar Energy 84*(9): 1717–1744.

Levinson R., Akbari H., Berdahl P., 2010c. Measuring solar reflectance – Part II: Review of practical methods. *Solar Energy 84*(9): 1745–1759.

Levinson R., Berdahl P., Berhe A., Akbari H., 2005b. Effects of soiling and cleaning on the reflectance and solar heat gain of a light-colored roofing membrane. *Journal of Atmospheric Environment 39*: 7807–7824.

Levinson R., Akbari H., Konopacki S., Bretz S., 2005a. Inclusion of cool roofs in non-residential Title 24 prescriptive requirements. *Energy Policy 33*(2): 151–170.

Levinson R., Akbari H., Reilly J.C., 2007. Cooler tile-roofed buildings with near-infrared-reflective non-white coatings. *Building and Environment 42*(7): 2591–2605.

Magos T., Plassais A., Bartzis J.G., Vasilakos C., Moussiopoulos N., Bonafous L., 2008. Photocatalytic degradation of NO_x in a pilot street canyon configuration using TiO_2-mortar panels. *Environmental Monitoring and Assessment 136*(1): 35–44.

Marchesi S., Zauli-Sajani S., Lauriola P., 2014. Quantitative evaluation of the impact of mitigation strategies for the urban heat island within the urban area of Modena. In *Proceedings of Third International Conference on Countermeasures to Urban Heat Island*. Venice, Italy. October 13–15.

Mastrapostoli E., Santamouris M., Kolokotsa D., Vassilis P., Venieri D., Gompakis K., 2014. A numerical and experimental analysis of the aging of the cool roofs for buildings in Greece, in proceedings of Third International Conference on Countermeasures to Urban Heat Island. Venice, Italy. October 13–15.

Menon S., Akbari H., Mahanama S., Sednev I., Levinson R., 2010. Radiative forcing and temperature response to changes in urban albedos and associated CO_2 offsets. *Environmental Research Letters 5*: 1–11.

Miller W.A., Desjarlais A., Atchley J., Keyhani M., MacDonald W., Olson R., Vandewater J., 2005. Experimental analysis of the natural convection effects observed within the closed cavity of tile roof systems. In *Proceedings of Cool Roofing . . . Cutting through the Glare*. Atlanta, GA. May 12–1.

Oleson K.W., Bonan G.B., Feddema J., 2010. Effects of white roofs on urban temperature in a global climate model. *Geophysical Research Letters 37*: L03701, 1–7.

ORNL SRI Calculator. Available online: http://www.ornl.gov/sci/roofs+walls/calculators/sreflect/index.htm. Accessed December 2014.

Paolini R., Zinzi M., Poli T., Carnielo E., Mainini A. G., 2014. Effect of ageing on solar spectral reflectance of roofing membranes: Natural exposure in Roma and Milano and the impact on the energy needs of commercial buildings. *Energy and Buildings 84*: 333–343.

Picada Project. Deliverable 21: Innovative façade coatings with de-soiling and de-polluting properties. Available online: www.picada-project.com. Accessed December 2014.

Pisello A. L., Cotana F., Brinchia L., 2014. On a cool coating for roof clay tiles: development of the prototype and thermal-energy assessment. *Energy Procedia 45*: 453–462

Poon C.S. and Cheung E., 2006. NO removal efficiency of photocatalytic paving blocks prepared with recycled materials. *Construction and Building Materials* 21(8):1746–1753.

Qin J., Song J., Qu J., Xue X., Zhang W., Song Z., Shi Y., Jiang L., Li J., Zhang T., 2014. The methods for creating building energy efficient cool black coatings. *Energy and Buildings* 84: 308–315.

Revel G.M., Martarelli M., Bengochea M.Á., Gozalbo A., Orts M.J., Gaki A., Gregou M., Taxiarchou M., Bianchin A., Emiliani M., 2013. Nanobased coatings with improved NIR reflecting properties for building envelope materials: Development and natural aging effect measurement. *Cement and Concrete Composites* 36: 128–135.

Revel G.M., Martarelli M., Emiliani M., Gozalbo A., Orts M.J., Bengochea M.Á., Delgado L.G., Gaki A., Katsiapi A., Taxiarchou M., Arabatzis I., Fasaki I., Hermanns S., 2014. Cool products for building envelope – Part I: Development and lab scale testing. *Solar Energy* 105: 780–791.

Romeo C., Zinzi M., 2013. Impact of a cool roof application on the energy and comfort performance in an existing non-residential building. A Sicilian case study. *Energy and Buildings* 67: 647–657.

Roof Savings Calculator by ORNL. Available online: http://www.roofcalc.com/. Accessed December 2014.

Rosenfeld A., Akbari H., Bretz S., Fishman B., Kurn D., Sailor D., Taha H., 1995. Mitigation of urban heat islands: Material, utility programs, updates. *Energy and Buildings* 22: 255–265.

Rosenfeld J., Romm J., Akbari H., Pomerantz M., 1998. Cool communities: Strategies for heat islands mitigation and smog reduction. *Energy and Buildings* 28: 51–62.

Rossi F., Pisello A.L., Nicolini A., Filipponi M., Palombo M., 2014. Analysis of retro-reflective surfaces for urban heat island mitigation: A new analytical model. *Applied Energy* 114: 621–631.

Sailor D.J., 1995. Simulated urban climate response to modifications in surface albedo and vegetative cover. *Journal of Applied Meteorology* 34(7): 1694–1704.

Santamouris M., Synnefa A., Kolokotsa D., Dimitriou V., Apostolakis K., 2008. Passive Cooling of the Built Environment-Use of Innovative Reflective Materials to Fight Heat Island and Decrease Cooling Needs. *International Journal of Low Carbon Technologies* 3(2): 71–82.

Sleiman M., Kirchstetter T. W., Berdahl P., Gilbert H. E., Quelen S., Marlot L., Preble C. V., Chen S., Montalbano A., Rossele O., Akbari H., Levinson R., Destaillats H., 2014. Soiling of building envelope surfaces and its effect on solar reflectance – Part II: Development of an accelerated aging method for roofing materials. *Solar Energy Materials and Solar Cells* 122: 271–281.

Sleiman M., Weiss G.B., Gilbert H.E., Francois D., Berdahl P., Kirchstetter T.W., Destaillats H., Levinson R., 2011. Soiling of building envelope surfaces and its effect on solar reflectance – Part I: Analysis of roofing product databases. *Solar Energy Materials and Solar Cells* 95: 3385–3399.

Stathopoulou M., Synnefa A., Cartalis C., Santamouris S., Karlessi T., Akbari H., 2009. A surface heat island study of Athens using high-resolution satellite imagery and measurements of the optical and thermal properties of commonly used building and paving materials. *International Journal of Sustainable Energy* 28: 59–76.

Strini A., Cassese S., Schiavi L., 2005. Measurement of benzene, toluene, ethylbenzene and oxylene gas phase photodegradation by titanium dioxide dispersed in cementitious materials using a mixed flow reactor. *Applied Catalysis B: Environmental* 61: 90–97.

Sugrañez R., Álvarez J.I., Cruz-Yusta M., Mármol I., Morales J., Vila J., Sánchez L., 2013. Enhanced photocatalytic degradation of NO_x gases by regulating the microstructure of mortar cement modified with titanium dioxide. *Building and Environment* 69: 55–63.

Synnefa A., Dandou A., Santamouris M., Tombrou M., Soulakellis N., 2008. On the use of cool materials as a heat island mitigation strategy. *Journal of Applied Meteorology and Climatology 47*: 2846–2856.

Synnefa A., Pantazaras A., Santamouris M., Bozonnet E., Doya M., Zinzi M., Muscio A., Libbra A., Ferrari C., Coccia V., Rossi F., Kolokotsa D., 2013. Interlaboratory comparison of cool roofing material measurement methods. In *Proceedings of Joint Conference 34th AIVC- 3rd TightVent – 2nd Cool Roofs' – 1st Venticool*. Athens, Greece. September 25–26.

Synnefa A., Saliari M., Santamouris M., 2012. Experimental and numerical assessment of the impact of increased roof reflectance on a school building in Athens. *Energy and Buildings 55*: 7–15.

Synnefa A., Santamouris M., Apostolakis K., 2007a. On the development, optical properties and thermal performance of cool colored coatings for the urban environment. *Solar Energy Journal 81*(4): 488–497.

Synnefa A., Santamouris M., Akbari H., 2007b. Estimating the effect of using cool coatings on energy load sand thermal comfort in residential buildings in various climatic conditions. *Energy and Buildings 39*(11): 1167–1174.

Synnefa A., Santamouris M., Livada I., 2006. A study of the thermal performance of reflective coatings for the urban environment. *Solar Energy Journal 80*: 968–981.

Taha H., 1997. Modeling the impacts of large-scale albedo changes on ozone air quality in the South Coast Air Basin. *Atmospheric Environment 31*: 1667–1676.

Taha H., 2005. *Urban surface modification as a potential ozone air-quality improvement strategy in California – Phase one: Initial mesoscale modeling*. Martinez, CA: Altostratus Inc. for the California Energy Commission, PIER Energy-Related Environmental Research. CEC-500–2005–128. Available online: http://www.energy.ca.gov/2005 publications/CEC-500–2005–128/CEC-500–2005–128.PDF. Accessed December 2014.

Taha H., 2007. *Urban surface modification as a potential ozone air-quality improvement strategy in California – Phase 2: Fine-resolution meteorological and photochemical modeling of urban heat islands*. Martinez, CA: Altostratus Inc. for the California Energy Commission, PIER Energy-Related Environmental Research.

Taha H., 2008a. Meso-urban meteorological and photochemical modeling of heat island mitigation. *Atmospheric Environment 42*(38): 8795–8809.

Taha H., 2008b. Urban surface modification as a potential ozone air-quality improvement strategy in California: A mesoscale modeling study. *Boundary-Layer Meteorology 127*: 219–239. doi: 10.1007/s10546-007-9259-5.

Taha H., Chang C., Akbari H., 2000. *Meteorological and air quality impacts of heat island mitigation measures in three U.S. cities*. LBL Report 44222. Berkeley, CA: Lawrence Berkeley National Laboratory.

Taha H., Hammer H., Akbari H., 2002. Meteorological and air quality impacts of increased urban surface albedo and vegetative cover in the Greater Toronto Area, Canada. LBNL Report 49210. Berkeley, CA: Lawrence Berkeley National Laboratory.

Taha H., Konopacki S., Gabersek S., 1999. Impacts of large-scale surface modifications on meteorological conditions and energy use: a 10-region modeling study. *Theoretical and Applied Climatology 62*(3–4): 175–185.

Taha H., Liu X, Sailor D., Meier A., Benjamin M., Winer A., Douglas S., Haney J., 1994. Analysis of energy efficiency and air quality in the South Coast Air Basin of California – Phase II. *LBL Report No. 35728, vol. 2.*

TX ACTIVE, 2006. TX ACTIVE: Presentation of the first active solution to the problem of pollution. Available online: http://www.italcementigroup.com/ENG/Media+and+Communication/News/Corporate+events/20060228.htm. Accessed December 2014.

Werle A.P., Lo K., John V.M., Ando R., de Souza M.L., 2014. Performance of self-cleaning cool cementitious surface. In *Proceedings of Third International Conference on Countermeasures to Urban Heat Island*. Venice, Italy. October 13–15.

Yalim M.S., Frère A., Goffaux C., 2014. CFD simulation of the three-dimensional effects induced by the NO_x photocatalytic degradation for isolated building and street-canyon configuration. In *Proceedings of Third International Conference on Countermeasures to Urban Heat Island*. Venice, Italy. October 13–15.

Zhang S.-H., Tanadi D., Li W., 2010. Effect of photocatalyst TiO_2 on workability, strength, and self-cleaning efficiency of mortars for applications in tropical environment. Presented at the *35th Conference on Our World in Concrete and Structures*. Singapore. August 26–27.

Zhang Y., Zhou G., Lin K., Zhang Q., Di H., 2007. Application of latent heat thermal energy storage in buildings: state of the art and outlook. *Building and Environment 42*, 2197–2209.

Zinzi M., 2014. Characterisation and assessment of near infrared reflective paintings for building façade applications. In *Proceedings of Third International Conference on Countermeasures to Urban Heat Island*. Venice, Italy. October 13–15.

5

COOL PAVEMENTS TO MITIGATE URBAN HEAT ISLANDS

Mat Santamouris

University of Athens – Athens, Greece

Introduction

Urban heat islands deal with the development of higher ambient temperatures in the urban zones compared to the surrounding suburban and rural areas (Santamouris, 2001). The specific characteristics as well as the magnitude of the urban heat island is highly determined by the urban layout, the type of materials used in the urban fabric, the type and characteristics of the pavements, the quantity of the anthropogenic heat released, the structural and morphological properties of the cities, the synoptic weather and climate conditions, possible local meteorological factors, etc. (Oke *et al.*, 1991). In parallel, it depends highly on the selection of the reference rural measuring station.

Mitigation of urban heat islands is a major task for scientists. Several mitigation technologies and systems are already proposed and used in real case studies with important results (Santamouris, 2014). In particular, the use of materials presenting a high reflectivity to the solar spectrum together with a high emissivity seems to present an important mitigation potential (Synnefa *et al.*, 2008).

Pavements in cities cover a very high part of the urban fabric. Pavement as a term involves paved surfaces like roads, sidewalks and parking areas, while most of the paved surfaces are made from concrete or asphalt. Statistics in the USA show that pavements occupy almost 29–39 per cent of the city's surface (Rose *et al.*, 2003; Akbari and Rose, 2001; Akbari *et al.*, 1999; Akbari *et al.*, 2001), while in the same country only the parking lots occupy 37.2 billion square meters covered by asphalt (Chester *et al.*, 2010). Pavements represent a high commercial interest, as just in the USA, improvements of the road surfaces, bridges and highways cost 166 billion dollars per year, while almost 80–90 billion dollars is spent for the preservation of the pavements (AASHTO, 2008).

The impact of urban pavements on the development of the urban heat island is quite important. Conventional paved surfaces absorb highly solar radiation and present a considerably high surface temperature. As a matter of fact, the sensible heat released to the atmosphere is quite high and contributes to increased ambient temperatures in the canopy layer. Several recent studies have shown, that conventional pavements are among the strong and important sources increasing the urban ambient temperature (Doulos *et al.*, 2004; Synnefa *et al.*, 2006). Thus, important research is organized around technologies able to decrease the surface temperature of pavements and the corresponding release of sensible heat. New pavement technologies may be used to replace existing pavements with new ones of higher mitigation efficiency or through the reconstruction, rehabilitation and preservation of the already installed pavements. In parallel, shading of paved surfaces can contribute highly to decreasing their surface temperature (Tran *et al.*, 2009).

Pavements developed to present a low surface temperature and mitigate urban heat island are known as 'cool pavements'. Two major clusters of available technologies may be identified as cool pavements: those that present a high reflectivity to the solar spectrum together with a high emissivity, known as reflective pavements, and those that are using the latent heat of the water evaporation to decrease their surface temperature, known as water retentive pavements. Both technologies are very well developed and the corresponding products are available in the market. Important projects have been already designed and built using both types of cool pavements and it is concluded that their mitigation potential is quite high. As a matter of fact, cool pavements have an important share in the market. In particular, only in Japan until 2009, the installed surface of cool pavements, was more than 800,000 m², while annual installations exceeded 270,000 m² (Takahashi, 2011).

Monitoring and simulation techniques

Simulation of the thermal performance of pavements may be performed using analytical or numerical techniques (Qin and Hiller, 2001; Solaimanian and Kennedy, 1993; Ramadhan and Al-Abdul Wahhab, 1997; Bentz, 2000; Gui *et al.*, 2007; Yavuzturk *et al.*, 2005; Hermansson 2004; Gui *et al.*, 2007). All proposed methods analyze the thermal balance of the pavements and use different types of numerical or analytical methods to solve the corresponding equations. Comparison against experimental data shows that most of the proposed models present an acceptable prediction accuracy.

Monitoring of the thermal performance of pavements may be performed using two specific techniques:

a) Mesoscale remote sensing techniques. These techniques are satellite based and have been used many times to monitor the surface temperature of paved surfaces (Golden and Kaloush, 2006; Barring *et al.*, 1985; Quattrochi and Ridd, 1994; Streutker, 2002; Streutker, 2003; Carnahan and Larson, 1990). Satellite monitoring of the thermal performance of pavements provide an excellent overview of the spatial distribution of the pavement surface temperature; however, they present several constraints like the time limitations set by the orbit of the satellites, the resolution of the images and the problems associated to the interpretation of the images.

b) Microscale measurement methods involving infrared thermography and conventional temperature monitoring. Use of infrared thermography requires a good knowledge of the pavements emissivity; however, this is simple to be done. The use of conventional measuring techniques based on the use of thermocouples requires a very good contact between the measuring devices and the surface. Microscale measuring techniques present, in general, a very high accuracy and provide results at any time without being able to visualize the surface temperature distribution in a large area. Golden and Kaloush (2006) have performed comparative measurements of the pavement surface temperature using both infrared cameras and thermocouple sensors. The temperature difference between the two methods ranged between 0.2 to 5.6 K, while higher temperatures are given by the infrared camera.

The physics of pavements

As already mentioned, pavements affect highly the urban climate and may be considered as the major contributors to the development of urban heat island (Asaeda *et al.*, 1996). Asaeda *et al.* (1996) have reported that in Tokyo, Japan, the emitted infrared radiation causes heating of the urban atmosphere equal to half of the energy consumption of the commercial areas in the city.

The thermal balance of pavements depends on the solar radiation absorbed, the emitted infrared radiation, the heat transfer through convection from and to the ambient air, heat losses and gains with the ground and the other solid surfaces, and also on the storage of heat in the mass of the surface. In case latent heat phenomena are present, the possible evaporation or condensation affects highly the thermal balance of pavements, while special phenomena like rain and icing have an important impact. In the following paragraph, the main optical and thermal parameters affecting the thermal behavior of pavements are analysed.

Thermal conductivity controls the conductive transfer of heat in pavements. The thermal conductivity of most of materials used for pavements is well known and documented, however the specific way that affects the thermal balance of pavements is not always understood. High thermal conductivity values are associated with a faster transfer of heat from and to the ground. When, during the daytime, the temperature of the pavement may be higher than that of the ground, heat is transferred from the pavement to the soil, while during the night period the adverse heat flow may happen. Thus, pavements having a high conductivity present a lower average maximum and a higher average minimum temperature. Gui *et al.* (2007) reported that when the thermal conductivity of pavements increases from 0.60 to 2.60 W/m/K, the average maximum pavement temperature decreases by 7 K, while the average minimum temperature increases by 4.5 K. However, experiments carried out in highly cloudy climates and reported by Hermansson (2004) concluded that thermal conductivity has a quite marginal role in determing the pavement surface temperature.

Thermal capacitance of the pavements defines the ability of the material to store heat in its mass. Pavements with a high thermal capacitance present, in general, a lower maximum and a higher minimum surface temperature. As reported by Gui *et al.* (2007) when the capacitance increases from 1.40 to 2.80 x10^6 J m^{-3} °C^{-1}, the average maximum temperature is decreasing by 5 K.

Absorbed solar radiation is the major source increasing the surface temperature of pavements. The amount of the absorbed solar radiation is controlled by the spectral and broadband absorptivity of the concerned material. Information on the broadband reflectivity of various pavements materials is given in Santamouris (2001). Reflectivity of the materials in the visible part of solar radiation depends mainly on its color and roughness. It well known that light colors present a higher reflectivity in the visible part of solar radiation than dark colors. It has to be clear that reflectivity in the infrared part of solar radiation is almost independent of the materials color. The roughness of the materials is also affecting the global absorptivity of pavements. Pavements with smooth and flat surfaces are generally cooler than similar materials with rough and anaglyph surface (Doulos *et al.*, 2004).

Various works are published, aiming to identify the relation between the pavements optical characteristics and the corresponding surface temperature.

Measurements reported by Niachou et al. (2008) show that the surface temperature of black asphalt during summer in Athens, Greece was 65°C, of the gray stone 48°C, and of the shaded concrete pavement 30°C. Taha et al. (1992), reported that white elastomeric coatings having albedo around 0.72, were almost 45 K cooler than black coatings having an albedo of 0.08. Doulos et al. (2004) has compared the surface temperature of various commercial pavements and reported that under highly sunny conditions, the maximum temperature difference between white marbles and black granites was close to 19 K. Synnefa et al. (2011) reported the surface temperature of various thin layer asphaltic materials measured during the summer period in Athens, Greece. While light color asphalt presented a broadband reflectivity equal to 0.45, the reflectivity of black asphalt was 0.03. In this case, the surface temperature difference between the two materials was close to 12 K. Other color asphalt layers like yellow, beige and green had a broadband reflectivity of 0.26, 0.31 and 0.1 and presented 9.0, 7.0 and 5.0 K lower surface temperature than the black asphalt. Using satellite based measurements, Stathopoulou et al. (2009) measured the surface temperature of various pavement materials in Athens. He found that black asphalt presented temperature around 70–80°C, while concrete was between 56 to 79°C, marbles between 49 to 67°C and stones between 47 to 75°C. Golden and Kaloush (2006) reported comparative measurements of different paving materials like conventional graded hot mix asphalt, asphalt rubber chip seal, gap-graded asphalt rubber mixture, plain concrete pavement sections and plain concrete sections modified with the use of crumb rubber. It is reported that the gap-graded asphalt rubber concrete presented the a surface temperature of 67.8°C that was the highest one, while its albedo was the lowest one (0.12). Concrete sections presented the highest albedo (0.48), while their surface temperature was 51.8°C. Also, the albedo of the thick asphalt rubber was 0.13 and its surface temperature was 66.7°C. When the albedo increased to 0.26 by adding some white paint, the surface temperature was reduced to 51.1°C. Pomerantz et al. (1997) has also measured the surface temperature of various paving materials with different albedos. It is reported that pavements having albedos of 0.05, 0.15 and 0.35 presented surface temperatures equal to 50.5, 46.1 and 32.2°C respectively. Gui et al. (2007) have performed a sensitivity analysis concerning the influence of the materials' albedo on their maximum and minimum surface temperature. It was reported that materials having albedos of 0.1 and 0.5 presented surface temperatures 71°C and 53°C, respectively.

Berg and Quinn (1978) performed comparative surface temperature measurements in streets having an albedo close to 0.55 and 0.15 and found that the ambient temperature was almost 11 K lower in the streets of higher albedo. In fact, the sensible heat transferred to the ambient atmosphere from pavements of high temperature is quite high. As mentioned by Asaeda et al. (1996), the sensible heat transferred by black asphalt is almost 200 W/m² higher than that of bare soil, while at the same time the emitted infrared radiation was also 150 W/m² higher.

Important research is carried out in order to increase the spectral and broadband reflectivity of the paving materials. Research efforts focus towards two specific directions: a) to increase the broadband reflectivity of white or light colored pavements by improving their spectral reflectance in the visible part of the solar radiation, and b) to improve the broadband reflectivity by increasing the spectral reflectivity of colored paving materials in the near infrared part of the solar spectrum.

Emitted infrared radiation by pavements varies as a function of their surface temperature and emissivity. The higher the emissivity of the materials, the higher the radiation emitted and the faster the release of the stored heat. In reality, the emissivity is the most important parameter controlling the pavement surface temperature during the nighttime (Synnefa et al 2006). Paved surfaces with an emissivity value of 0.93 were found to present an almost 5 K lower surface temperature during the nighttime compared to pavements with an emissivity around 0.3 (Synnefa et al., 2006). In a similar way, Gui et al. (2007) found that when the emissivity value increases from 0.7 to 1.0, the maximum surface temperature of pavements decreases by 5 K, while the decrease of the minimum temperature is 8.5 K. However, Oke et al. (1991) reported that the impact of the materials' emissivity on the nighttime intensity of heat island is quite minor. When the emissivity increased from 0.85 to 1.0, the intensity of the UHI varied by 0.4 K, and just for very narrow canyons. The absolute value of the infrared radiation over asphalt and concrete is measured by Asaeda et al. (1996) and is found that maximum upward radiation was 550 W/m², while the net infrared radiation balance was between 0.0 to 80 W/m². Measurements reported by Qin and Hiller (2001) show that the net infrared from pavements varied between 60 to 120 W/m².

The combined impact of reflectivity to solar radiation and thermal emissivity of the pavements materials on their surface temperature is studied by Shi and Zhang (2011). It is found that the impact of the emissivity becomes very important when the reflectivity is low, while for higher values of the reflectivity the impact of emissivity is reduced. In study of similar nature, Gui et al. (2007) found that both the solar reflectivity and emissivity of the pavements have the highest positive impact on the surface temperature of the materials.

Convective heat transfer is a very important term in the thermal balance of pavements. Convective gains or losses depend on the temperature difference between the pavement and the ambient air as well as on the value of the heat transfer coefficient. The latest depends on the type of flow, laminar or forced, and in general, it is a function of the wind speed. The magnitude of convective transfer over asphalt and concrete pavements is measured by Asaeda et al. (1996), and it is found that the maximum convective heat transfer during the hottest time of the day was 350 and 200 W/m² respectively.

Technologies of cool pavements

The main objective of cool pavements is to reduce the sensible heat released to the atmosphere. This can be achieved by reducing mainly the surface temperature of the paved surfaces. There are different technologies and methods to decrease the surface temperature of pavements. For sure, the more known, and perhaps the more effective one, is the shading of the paving surfaces. Other known and used techniques are those employing materials presenting a high reflectivity to the solar radiation together with a high emissivity, and also those presenting a high permeability that allow storage of water on their mass, which is used to decrease their surface temperature through evaporation. In the following, four technologies for cool pavements are presented and discussed.

a) Reflective pavements are those that use surface materials of high solar reflectivity and emissivity to decrease the absorbed solar radiation and increase the emitted infrared radiation. Reflective pavement techniques are mainly applied to concrete and asphalt and may be used for new pavements or for pavement rehabilitation and maintenance, Reflective pavement technologies involve various techniques like the application of white topping and ultra-thin white topping techniques, the application of color pigments and seals, the use of colorless and reflective synthetic binders, the use of concrete additives like slag cement and fly ash, the use of light aggregates in asphalt concrete surfaces, the use of roller compacted concrete pavement, the use of chip or sand seals with light aggregates, the painting of the surfaces with a light color, using or not using microsurfacing techniques, the use of sand/shot blasting, and abrading binder surfaces, resin based pavements, etc. (EPA 2005, Tran and Powell 2009, Pomerantz et al., 1997).

b) Permeable, porous, pervious or water-retaining pavements present a high permeability and decrease their surface temperature through evaporation. This type of pavement presents a lower albedo than impermeable surfaces because generally they are characterised by a higher effective surface created by the higher void content (Haselbach 2009). As a matter of fact, permeable pavements present higher convective fluxes to the atmosphere than impermeable pavements. Permeable pavements may be vegetable or not, and some of the more common techniques are porous and pervious concrete, porous or rubberized asphalt, concrete and plastic grid pavers filled with gravels, and permeable interlocking concrete pavers, (EPA 2005, Tran and Powell 2009, Pomerantz et al., 1997). Some of the more known vegetable techniques involve concrete grid pavers, the use of lattices of different types that allow grass to grow in the interstices, grass pavers, and reinforced turf.

c) Pavements of increased thermal capacitance. Given that paving materials present a high thermal capacitance, usually phase-change materials are added to increase their storage capacity.

d) Pavements linked to mechanical systems. This involve flow of water in the mass of the pavements, underground circulation of wastewaters, etc.

In the following paragraph, the main technologies and characteristics of the above paving technologies are presented.

Reflective pavements

Reflective pavements present a high reflectivity to the solar spectrum and this results in a much lower surface temperature and a reduced sensible heat released to the ambient atmosphere. In parallel, reflective pavements decrease the need for artificial light during the night. According to Pomerantz et al. (2000), the reflectivity of the pavements may increase either by covering the surface of the pavement with a reflective coating, by using aggregates of light color, or by using an appropriate binder or by combining the above techniques.

Many existing techniques are available to increase the albedo of pavements. A comparative study of eight commercial technologies to increase the albedo of asphalt pavements is presented by Tran et al. (2009). As reported, six of the above technologies present an SRI value higher than 0.29. Chip and sand seal is a technique where the binder is first sprayed on the pavement, then on top of it the aggregates are placed and pressed. According to the existing knowledge, chip seal may help to decrease the surface temperature of asphalt pavements by 9.0 K. In this case, the binder and the aggregates are first mixed and then are placed on the pavement, the corresponding technique may be known as microsurfacing (addition of polymers), fog coating, slurry coating, etc. (Pomerantz et al., 2000, Hunter 1994). Color additives may be used as well to increase the albedo of the pavements. White topping techniques are based on the use of concrete overlays on the surface of asphalt pavements. New concrete presents three to four times higher albedo than new asphalt and thus, white topping techniques contribute to increase highly the albedo of the overall pavement. The thickness of white toppings may vary between 5 to 20 cm, as a function of the used technique.

Important research is actually carried out in order to develop more advanced reflective pavements. According to Santamouris (2013), five are the main research directions on this topic (Figure 5.1):

a) The use of high reflective paints on the visible solar spectrum applied on the surface of the pavement: These ultra-white paints may present a reflectivity higher than 90 per cent. New generation white paints of new generation present a very high solar reflectivity that in many cases exceeds 90 per cent (Santamouris *et al.*, 2011).

As expected, the application of high reflective paints on pavements decreases their surface temperature and the sensible heat released to the atmosphere. Experimental results from similar applications during the summer period are reported by Santamouris *et al.* (2008) and Synnefa *et al.* (2006). The albedo of the measured pavements was always higher than 0.8 and lower than 0.9 and the reflective pavements were compared against conventional white, paved surfaces. Synnefa *et al.* (2006), measured and compared the surface temperature of 14 highly reflective white paints available in the market.

A part of the pavements covered by aluminum paints presented an emissivity between 0.3–0.4; all other coatings had an emissivity close to 0.8.

The reported reduction of the surface temperature of the measured concrete pavements was 4 K during the daytime and 2 K during the night, while the pavements covered with aluminum paints presented the higher surface temperature, and the acrylic elastomeric the lower one. Compared to the ambient temperature, pavements were 2 K warmer during the day and almost 6 K cooler during the night period.

In another experiment described by Santamouris *et al.* (2008), low-cost coatings based on the use of calcium hydroxide were applied on concrete pavements and tested in comparison against conventional white color paved surfaces. The albedo of the tested pavement was 0.88, while the albedo of the conventional one was 0.76. During the daytime, the tested pavement presented 1–5 K lower surface temperature than the conventional one.

b) Use of infrared reflective coatings on the surface of the pavements: These non-white coatings present a higher solar reflectivity in the infrared that the conventional coatings of the same color. Several experiments are carried out aiming to identify the efficiency of the infrared reflective coating when applied on the surface of concrete and asphalt pavements. Synnefa *et al.*, 2007, describes the results of an experiment where ten infrared reflective coatings of different colors were applied on concrete pavements and tested during the summer period. Black color pavements with infrared reflective coatings presented an albedo close to 0.27 while the albedo of the conventional black pavement was 0.05. In this case, the measured decrease of the surface temperature was almost 10 K. Also the reflective blue pavement has an albedo of 0.33 against 0.18 of the conventional blue, while the surface temperature difference was 4.5 K. Similar results are also reported by Kondo *et al.* (2008). In another experiment reported by Wan *et al.* (2009), dark color infrared reflective coatings are used on the surface of concrete pavements. A surface temperature difference of about 5 K is measured between the reflective and the conventional pavement. In similar experiments, reported Belkowitz (2011) and Levinson (2011), surface temperature differences between 11–22 K are reported. Synnefa *et al.*, 2011, reported the use of infrared reflective coatings on asphalt pavements. In particular, infrared reflective paints were mixed with colorless elastomeric asphalt binders to develop five thin-layer asphalt pavements. The albedo of the reflective asphalt pavement varied between 27 per cent to 55 per cent, while the albedo of the conventional asphaltic pavement was 4 per cent. During the day period, the surface temperature of the reflective asphaltic pavement was between 36°C to 44°C, while the corresponding temperature of the conventional pavement was 60°C. Also, during the night period, reflective pavements presented 1–2 K of lower surface temperature.

c) **Use of reflective paints to cover the aggregates of asphaltic pavements**: Such a technology is proposed and tested by Boriboonsomsin and Farhad (2011) and Kawakami and Kubo (2008). Boriboonsomsin and Farhad (2011) have coated aggregates with a heat reflecting paint. The final albedo of the pavement was between 0.46 to 0.57. Such a pavement presented a surface temperature difference between 10.2 K and 18.8 K against conventional asphaltic surfaces. Kawakami and Kubo (2008) used an almost similar technology and they found that when the albedo of the surface increases by 0.25, the surface temperature increases by 6.8 K against a conventional pavement. For an albedo increase of 0.6 the surface temperature decreased by about 20 K.

d) **Use of color-changing coatings applied on the surface of the pavements**: Such a technology is proposed by Karlessi *et al.* (2011). Specifically, thermochromic coatings were used on the surface of the pavements. Such coatings are able to change color to respond to the specific ambient temperature. During the warm summer period, the coatings become white, while during the winter period, they are dark. Eleven different concrete pavements were prepared and tested during the summer period Karlessi *et al.* (2011). It is found that the average surface temperature of the thermochromic pavements was almost 7 K lower

Figure 5.1
Current R + D directions on reflective pavements.
Source: Adapted from Santamouris, M.: Using cool pavements as a mitigation strategy to fight urban heat island – A review of the actual developments. *Renewable and Sustainable Energy Reviews.* Volume 26, October 2013, Pages 224–240

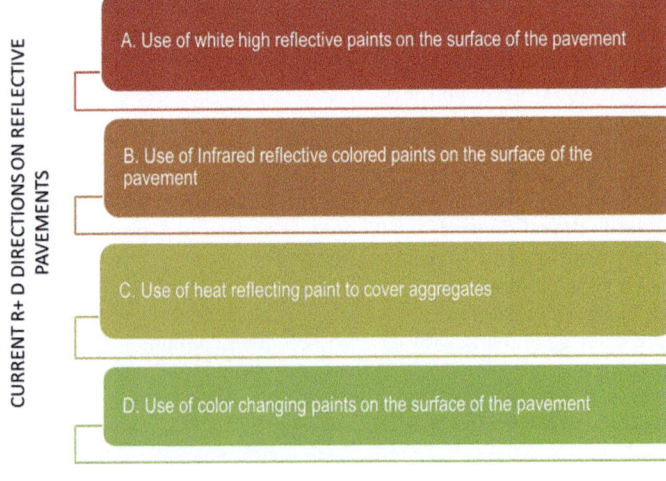

that of infrared reflective pavements and almost 10 K lower than the surface temperature of conventional pavements. The problem of ageing appeared to be the major problem of these type of reflective pavements.

e) Use of fly ash and slag as constituents of the concrete: Such a technology is proposed by Boriboonsomsin and Farhad (2011). It is reported that when slag is used to replace part of the concrete, the albedo of the pavement increased by 71 per cent.

Permeable pavements

Permeable pavements are composed by voids, pores or water-holding fillers that allow water to pass and be stored in the mass of the material. Permeable pavements can be vegetated or not. When the temperature of the material increases, then water evaporates and cools the pavement. Evaporation rates depend strongly on the water content in the material, the temperature of the pavement and the atmospheric humidity. Comparison of the surface temperature of permeable pavements against conventional ones, shown that when permeable pavements are dry, their surface temperature is higher (Haselbach, 2009). In parallel, research carried out by Karasawa et al. (2006) has shown that no correlation exists between the permeability of permeable pavements and their surface temperature. Permeable pavements are more suitable for humid climates where availability of water is not a problem. In many cases wastewater may be used when available. The main advantages and disadvantages of permeable pavements are summarized by Scholz and Grabowiecki (2007). Three are the performance criteria for permeable pavements (Takahashi and Yabuta, 2009): a) how the material decreases its surface temperature under fine weather, b) the sustainability to suppress the temperature rise after rainfalls and c) the durability and the possible decrease of its performance over time.

Vegetative pavements are mainly composed by permeable interlocking units filled with soil and planted at the top with grass or other plants. Evaporation for the planted surface helps to achieve cooling. The surface temperature of the plants is usually lower tha. the surface of concrete or asphaltic materials used in pavements (Takebayashi and Moriyama, 2012). Permeable pavements based on concrete are using cement, aggregates and water and some additional materials like silica fumes, fly ash, pozzolans and ground blast furnace slag. Concrete permeable pavements have pores in the cement past, the aggregate voids and the air voids with the last one being the more important regarding the water permeability. Permeable pavements using asphaltic products involve the use of fine and coarse stone aggregates bound by a bituminous-based binder. The specific reflectivity of permeable asphaltic pavements depends on the reflectivity of the individual materials used.

As in the case of reflecting materials, important research is carried out on the field of permeable pavements. According to Santamouris (2013), six are the main research technological approaches on this field (Figure 5.2).

The main technologies are:

a) *Use of water-holding fillers made of steel byproducts as an additive to porous asphalt*: Nakayama and Tsuyoshi (2010) have proposed a permeable pavement including water-holding fillers made by steel byproducts and integrated in porous asphalt. As reported, the mean surface temperature of

Figure 5.2
Current R + D directions on permeable pavements.
Source: Adapted from Santamouris, M. Using cool pavements as a mitigation strategy to fight urban heat island—A review of the actual developments. *Renewable and Sustainable Energy Reviews*. Volume 26, October 2013, Pages 224–240

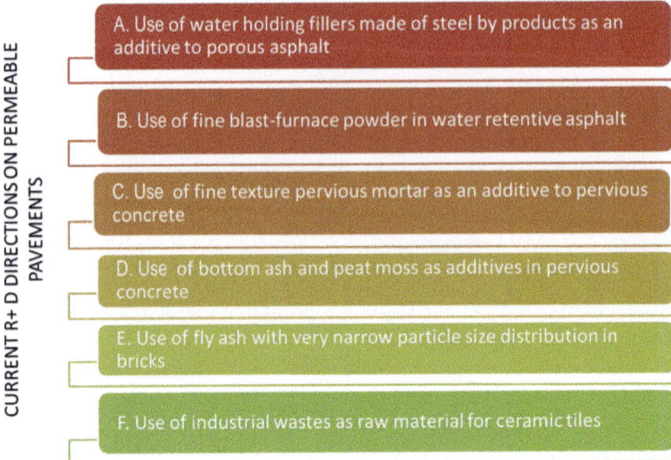

CURRENT R+ D DIRECTIONS ON PERMEABLE PAVEMENTS

A. Use of water holding fillers made of steel by products as an additive to porous asphalt

B. Use of fine blast-furnace powder in water retentive asphalt

C. Use of fine texture pervious mortar as an additive to pervious concrete

D. Use of bottom ash and peat moss as additives in pervious concrete

E. Use of fly ash with very narrow particle size distribution in bricks

F. Use of industrial wastes as raw material for ceramic tiles

the specific permeable pavement was 0.6 K lower than that of the conventional porous asphalt.

b) *Use of fine blast-furnace powder in water-retentive asphalt*. Takahashi and Yabuta (2009) have presented and tested a new permeable pavement material using fine blast-furnace powder. Such a material is produced through a blast furnace process. Testing of the pavement material showed that its surface temperature was almost 14°K lower than that of a dense graded asphalt pavement.

c) *Use of fine-texture pervious mortar as an additive to pervious concrete*. Yukari (2009) has developed and tested a new permeable concrete pavement combined using fine-texture pervious mortar. The mortar was produced using cementitious materials, aggregate, and water.

d) *Use of bottom ash and peat moss as additives in pervious concrete*. Park has developed and tested a new permeable pavement using bottom ash and peat moss. Testing of the new pavement showed that it presented almost 18°K lower surface temperature than the asphalt after a rainfall, while the maximum difference against a conventional permeable pavement was almost 9°K.

e) *Use of fly ash with very narrow particle size distribution in bricks*. Yokota *et al.* (2010) have developed new porous bricks with a porosity of 22–43 per cent. Testing shows that this type of pavement improves thermal comfort.

f) *Use of industrial wastes as raw material for ceramic tiles*. Tetsuji and Suzuki (1998) have developed and tested a novel permeable ceramic pavement. This pavement was developed using industrial wastes as raw material. Testing has shown that under saturation conditions the pavement was almost 10°K of lower temperature than a concrete roof, while the air above the pavement was 1–2°K of lower temperature.

Other mitigation pavement technologies

In the previous sections, the main developments on the field of reflective and permeable pavements were presented. However, some more technologies are proposed, tested and, in some cases, applied in real projects. Three are the main technologies proposed, as presented in Figure 5.3.

CURRENT R+ D DIRECTIONS ON OTHER PAVEMENT TECHNOLOGIES

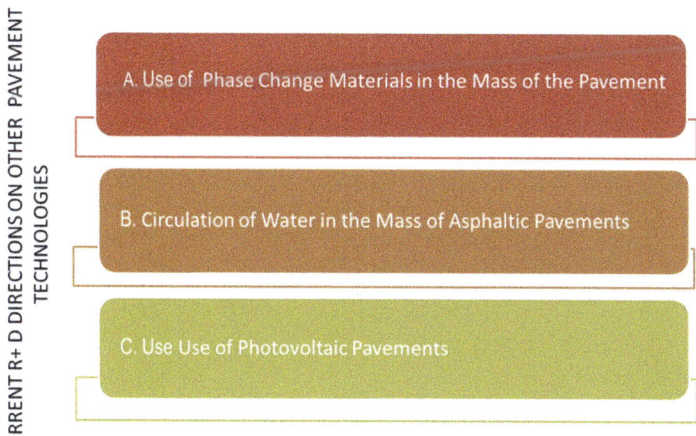

A. Use of Phase Change Materials in the Mass of the Pavement

B. Circulation of Water in the Mass of Asphaltic Pavements

C. Use Use of Photovoltaic Pavements

Figure 5.3
Current R + D directions on other pavements technology.
Source: Adapted from Santamouris, M. Using cool pavements as a mitigation strategy to fight urban heat island—A review of the actual developments. *Renewable and Sustainable Energy Reviews*. Volume 26, October 2013, Pages 224–240

a) *Use of phase-change materials in the mass of the pavement*. One way to decrease the surface temperature of the pavements and also the sensible heat released to the atmosphere is to increase the thermal capacity of the pavements. This can be achieved by including phase-change materials in the mass of the pavement. Such a technology is proposed by Karlessi *et al.* (2011). In particular, infrared reflective pigments were used together with nano pcm components in concrete pavements to prepare surfaces of different colors.

Testing of the pavements during the summer period has shown that the PCM-doped pavements presented about 2.9 to 8.3°K lower surface temperatures compared to the conventional tiles of the same color, and also 0.6 to 2.6°K lower temperature compared to the infrared reflective pavements.

Figure 5.4
A photovoltaic pavement in Kobe, Japan.

b) *Circulation of water in the mass of asphaltic pavements*. This is a very common idea investigated by many researchers (Mallick *et al.* 2009). The principle is based on the circulation of water through pipes placed behind its surface. Testing and calculations carried out show that the cooling potential of the system is high, but several problems have to be solved, like the reduced heat transfer because of the low conductivity of the asphalt.

c) *Use of Photovoltaic Pavements*. Photovoltaic pavements may generate electricity, save space and can contribute to mitigating heat islands in cities. Experimental results obtained during the summer of 2012 in Athens show that the surface temperature of photovoltaic pavements may be 3–5°K lower than that of conventional concrete pavements (Figure 5.4).

Case studies

Several projects using advanced cool pavements are already designed, built and monitored. Some of the existing examples where full documentation is available are presented in this chapter. The examples are built in Greece and all of them are using reflective pavements. More information on examples using other types of pavements are given by Santamouris (2013).

The Flisvos project

The urban park of the Flisvos area is located in the southwestern part of Athens and covers a total area close to 80.000 m² (Figure 5.5). Before the rehabilitation of the area, pavements were made of asphalt, concrete and dark paving materials. The absorptivity of the pavements were between 0.55 to 0.65, while zones covered by concrete and asphalt had absorptivities between 0.79 and 0.89 respectively. Experimental investigations have shown that this area is among the city zones with the higher ambient temperature.

Figure 5.5 Aerial View of the Flisvos Park.
Source: Reprinted from Santamouris, M., Gaitani, N., Spanou, A., Saliari, M., Giannopoulou, K., and Vasilakopoulou, K. Kardomateas, T. (2012). Using cool paving materials to improve microclimate of urban areas – Design realization and results of the Flisvos project. *Building and Environment*, 53, 128–136. doi:10.1016/j.buildenv.2012.01.022 with permission from Elsevier.

The local authorities have undertaken a major refurbishment for the area, implemented in two phases. During the first stage, about 2,500 trees and bushes were planted, while during the second phase, almost 4,500 m² of existing pavements were replaced with reflective tiles. The solar reflectance of the selected cool tiles was around to 60 per cent, and the corresponding reflectivity in the visible and infrared were 47 and 71 per cent respectively. A detailed description of the project is given by Santamouris *et al.* (2012a).

The thermal performance of the park was monitored before and after the installation of the cool pavement, and specific simulations were carried out to calibrate the theoretical calculations with the corresponding measurements. The calculated spatial distributions of the ambient temperature with and without the cool pavements are given in Figure 5.6. It is found that the peak ambient temperature in the interior zones of the park is 1.9 K lower because of the cool pavement.

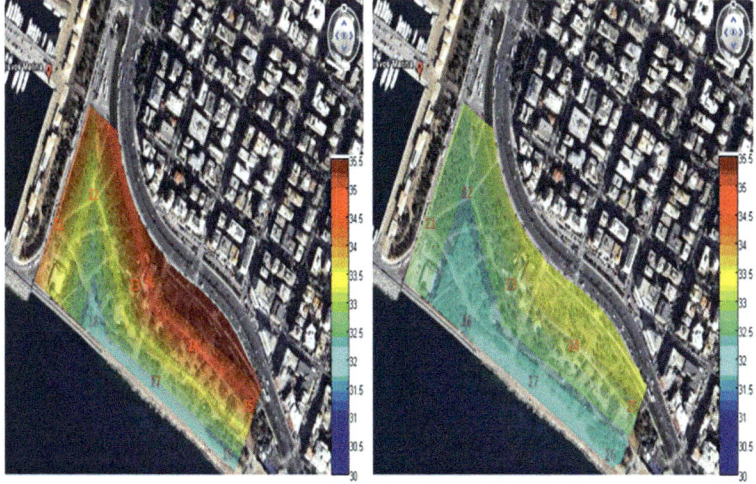

a. Without the Cool Pavement

b. With the Cool Pavement

Figure 5.6 Ambient temperature distribution in the park before and after the installation of the cool pavement.

Source: Reprinted from Santamouris, M., Gaitani, N., Spanou, A., Saliari, M., Giannopoulou, K., and Vasilakopoulou, K., Kardomateas, T. (2012). Using cool paving materials to improve microclimate of urban areas – Design realization and results of the Flisvos project. *Building and Environment*, 53, 128–136. doi:10.1016/j.buildenv.2012.01.022 with permission from Elsevier.

The Marousi case study

The considered project deals with the rehabilitation of an urban area located northeast of Athens. Prior monitoring of the ambient temperature has shown that the specific urban zone is characterized by high summer ambient temperatures.

The local authorities have decided to undertake a major rehabilitation of the area in order to improve its urban environmental quality. The total urban area rehabilitated was close to 16,000 m² and this covered a global zone of 277,000 m² (Figure 5.7). More information on the project may be obtained from Santamouris et al (2012b).

Before the rehabilitation most of the surfaces were covered with black asphalt and concrete tiles of quite dark color with a solar reflectivity not higher than 0.4. The area was monitored in details while simulations techniques were used to identify the possible benefits from the use of cool pavements. The project involved the following interventions: extended use of shading and solar control in the area, use of high size trees and pergolas to enhance shading and evapo-transpiration, use of reflective pavements with a reflectivity between 0.7 and 0.8 and use of earth-to-air heat exchangers in specific zones.

Figure 5.7
Aerial view of the Marousi area

Figure 5.8
Calculated distribution of
the ambient temperature in
the considered urban zone
for the proposed design
and the average peak
summer conditions

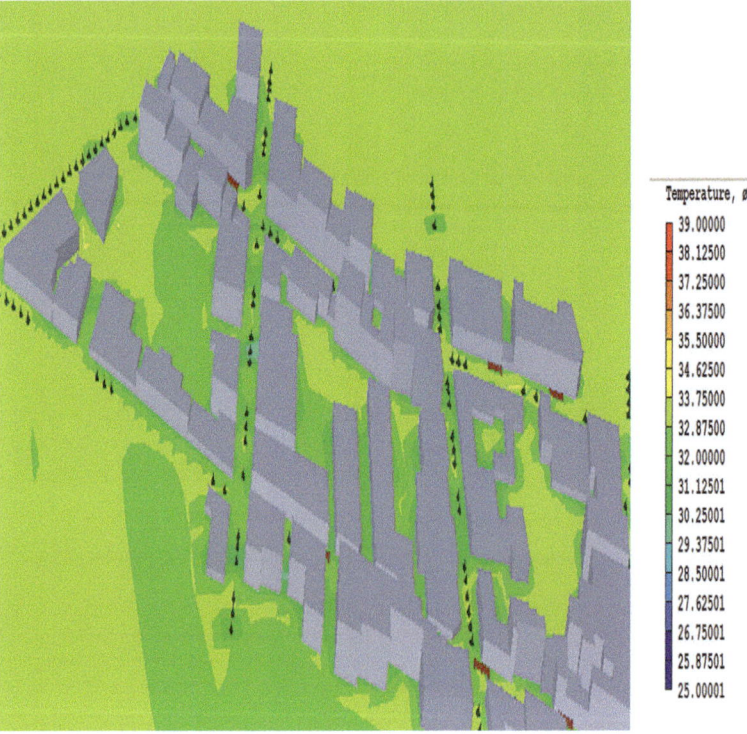

Figure 5.8 shows the calculated distribution of the ambient temperature in the considered urban zone after the installation of the cool pavements. Temperature refers to the average peak summer conditions. It is calculated that the reduction of the ambient temperature ranges between 2.3–3.4°C.

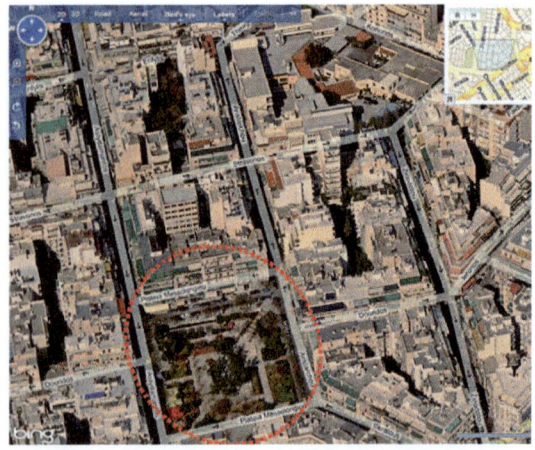

Figure 5.9 Satellite image of the area under rehabilitation.
Available at: http://www.bing.com/maps (accessed 1 July 2010).

A square in the center of Athens

This project deals with the rehabilitation of a central square in Athens using reflecting pavements and other mitigation technologies. The rehabilitated area had a total surface of 4160 m², (Figure 5.9). More information on this project is given by Gaitani *et al.* (2011).

The area was fully monitored to access the possible benefits of the application of the mitigation technologies, where extended and detailed simulations were also carried out. Reflective pavements were used together with additional green spaces and earth-to-air heat exchangers. The solar reflectivity of the pavements was 0.68.

Figure 5.10 gives the calculated distribution of the ambient temperature in the square for representative summer conditions. It is found that the ambient temperature in the area decreases by about 1–2°C compared to the existing situation.

Conclusions

Pavements play a very important role on the urban climate. High surface temperatures of the pavements increase the sensible heat released to the atmosphere and contribute highly to the development of the urban heat island. Intensive research carried out in recent years has permitted the development of efficient pavement technologies that contribute to lower urban temperatures. Important developments have been achieved on the field of reflective and permeable pavements. These new materials and techniques are extensively tested in laboratories and in real case studies, while important demonstration

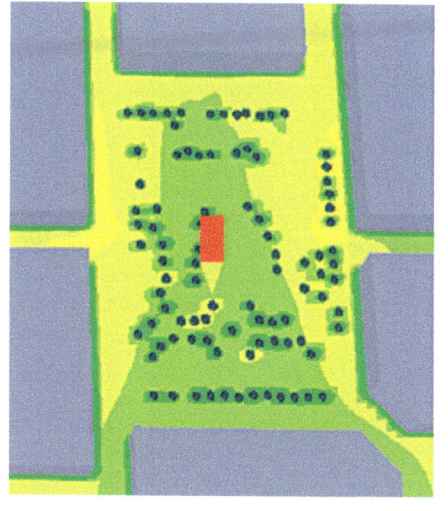

Figure 5.10 Calculated distribution of the ambient temperature in the square.

projects of a large scale are already built and monitored. Most of the monitored projects show that cool pavements present a much lower surface temperature than the conventional ones, while the peak ambient temperature in the area decreases up to 2 K.

Because of the high mitigation potential of the cool pavements, important industrial and market activities are undertaken and new products are entering the market continuously. It is evident that more focused research is necessary to be carried out in the near future. Products of higher efficiency have to be developed, while the real cooling potential of cool pavements has to be better documented.

References

American Association of State Highway and Transportation Officials (AASHTO). 2008. Guide for Pavement Friction. AASHTO Joint Technical Committee on Pavements, Washington, DC.

Akbari, H., Pomerantz, M. and Taha, H. (2001). Cool surfaces and shade trees to reduce energy use and improve air quality in urban areas. *Solar Energy*, 70(3), 295–310.

Akbari, H. and Rose, L.S. (2001). Characterizing the Fabric of the Urban Environment: A Case Study of Metropolitan Chicago, Illinois. (Paper LBNL-49275). Lawrence Berkeley National Laboratory, Berkeley, CA.

Akbari, H., Rose, L.S., and Taha, H. (1999). Characterizing the Fabric of the Urban Environment: A Case Study of Sacramento, California. (Paper LBNL-44688). Lawrence Berkeley National Laboratory, Berkeley, CA.

Asaeda, T., Thanh Ca, V. and Wake, A. (1996). Heat storage of pavement and its effect on the lower atmosphere. *Atmospheric Environment*, 30(3), 413–427.

Barring, L., Mattson, J. and Lindqvist, S. (1985)., Canyon geometry, street temperatures and urban heat island in Malmo, Sweden. *Int. J. Climatol.*, 5, 433–444.

Belkowitz, J. (2011, June). Can nanotechnology in concrete improve our roadways? Emerald Cities Sustainability Conference, Phoenix, AZ, USA.

Bentz, D.P. (2000). A computer model to predict the surface temperature and time-of-wetness of concrete pavements and bridge decks. (NISTIR 6551.) National Institute of standards and technology.

Berg, R. and Quinn, W. (1978). Use of light colored surface to reduce seasonal thaw penetration beneath embankments on permafrost. *Proceedings of the second international symposium on cold regions engineering, University of Alaska*, 86–99.

Boriboonsomsin, K. and Farhad, R. (2011). Mix design and benefit evaluation of high solar reflectance concrete for pavements. Transportation Research Record 0361–1981, 11–20.

Carnahan, W.H. and Larson, R.C. (1990). An analysis of an urban heat sink. *Remote Sensing of Environment*, 33, 65–71.

Chester, M., Horvath, A. and Madanat, S. (2010.) Parking infrastructure: energy, emissions, and automobile life-cycle environmental accounting. July 29, 2010. stacks.iop.org/ERL/5/034001

Doulos, L., Santamouris, M., and Livada, I. (2004). Passive Cooling of outdoor urban spaces. The role of materials. *Solar Energy*, 77(2), 231–249.

EPA. (2005)., Cool Pavement Report, EPA Cool Pavements Study – Task 5. Prepared for Heat Island Reduction Initiative, U.S. Environmental Protection Agency. Prepared by Cambridge Systematics, Inc. 4445 Willard Avenue, Suite 300 Chevy Chase, Maryland 20815.

Gaitani, N., Spanou, A., Saliari, M., Synnefa, A., Vassilakopoulou, K., Papadopoulou, K., Pavlou, K., Santamouris, M., Papaioannou, M., Lagoudaki, A. (2011). Improving the Microclimate in Urban Areas. A Case Study in the Centre of Athens, *Journal Building Services Engineers*, 32(1), 53–71.

Golden, J. S. and Kaloush, K. E. (2006). Mesoscale and microscale evaluation of surface pavement impacts on the urban heat island effects, *International Journal of Pavement Engineering*, 7(1), 37–52.

Gui, J., et al. (2007). Impact of pavement thermophysical properties on surface temperatures. *J Mater Civ Eng*, 19(8), 683–690.

Haselbach, L. (2009, Jan.). Pervious Concrete and Mitigation of the Urban Heat Island Effect. Transportation Research Board 88th Annual Meeting, Washington, D.C.

Hermansson, A. (2004). Mathematical model for paved surface summer and winter temperature: comparison of calculated and measured temperatures. *Cold Regions Science and Technology*, 40, 1–17.

Hunter, R. N. (Ed.). (1994). *Bituminous Mixtures in Road Construction*. Thomas Telford Ltd., London, UK.

Karasawa, A., Toriiminami, K., Ezumi, N., Kamaya, K. (2006, Nov.). Evaluation of Performance of Water-Retentive Concrete Block Pavements. 8th International Conference on Concrete Block Paving, San Francisco, CA.

Karlessi, T., Santamouris, M., Synnefa, A., Assimakopoulos, D., Didaskalopoulos, P., Apostolakis, K. (2011). Development and testing of PCM doped cool colored coatings to mitigate urban heat island and cool buildings. *Building and Environment*, 46(3), 570–576.

Kawakami, A. and Kubo, K. (2008). Development of a cool pavement for mitigating the urban heat island effect in Japan. In: 1st International symposium on asphalt pavements and environment, International Society for Asphalt Pavements, Zurich, Switzerland.

Kondo, Y., Ogasawara, T. and Kanamori, H. (2008). Field measurements and heat budget analysis on sensible heat flux from pavements. *J. Env. Engineering (AIJ)*, 73(628), 791–797.

Levinson, R. (2011, June). Cool Pavements for Cool Communities. Emerald Cities Sustainability Conference, Phoenix, Arizona.

Mallick, R., Chen, B.- L. and Bhowmick, S. (2009). Reduction of Urban Heat Island Effect through Harvest of Heat Energy from Asphalt Pavements. Proceedings of the Heat Island Conference, San Francisco, CA.

Nakayama, T. and Fujita, T. (2010). Cooling Effect of Water-Holding Pavements Made of New Materials on Water and Heat Budgets in Urban Areas. *Landscape and Urban Planning*, 96, 57–67. doi:10.1016/j.landurbplan.2010.02.003.

Niachou, K., Livada, I. and Santamouris, M. (2008). Experimental Study of Temperature and Airflow Distribution Inside an Urban Street Canyon during Hot Summer Weather Conditions. Part 1. Air and Surface Temperatures, *J. Buildings Environment*, 43(8), 1383–1392.

Oke, T.R, Johnson, G.T., Steyn, D.G. and Watson, I.D. (1991). Simulation of Surface Urban Heat Islands under 'Ideal' Conditions at Night – Part 2: Diagnosis and Causation. *Boundary Layer Meteorology* 56, 339–358.

Park, J.B., Kim, R.H., Lee, S.H., Pyun, H.B. (2010). Experimental Study on Thermal Characteristics of Environment-Friendly Blocks with Water-Absorbing Polymer. *Materials Science Forum*, 658, 260–263.

Pomerantz, M., Akbari, H., Chen, A., Taha, H., and Rosenfeld, A. H. (1997, Nov.). Paving Materials for Heat Island Mitigation, LBL-3 8074. The work was supported by the U.S. Environmental Protection Agency, and by the Office of Energy Efficiency and Renewable Energy, Office of Building Technologies, of the U.S. Department of Energy under Contract No. DE-ACO3–76SF00098, Heat Island Project, Environmental Energy Technologies Division, Ernest Orlando Lawrence Berkeley National Laboratory, University of California, Berkeley, CA 94720, U.S. Department of Energy, Washington, D.C. 20585.

Pomerantz, M., Pon, B., Akbari, H., and Chang, S. C. (2000). The effects of pavement temperatures on air temperatures in large cities. Heat Island Group (LBNL-43442.) Lawrence Berkeley National Laboratory, Berkeley, CA.

Qin, Y. and Hiller, J. E. (2001). Modeling temperature distribution in rigid pavement slabs: Impact of air temperature. *Construction and Building Materials*, 25, 3753–3761.

Quattrochi, D.A. and Ridd, M.K. (1994). Measurement and analysis of thermal energy responses from discrete urban surfaces using remote sensing data. *Int. J. Rem. Sens.*, 15, 10, 1991–2022.

Ramadhan, R. H. and Al-Abdul Wahhab, H. I. (1997). Temperature variation of flexible and rigid pavements in Eastern Saudi Arabia. *Building and Environment*, 32(4), 367–373.

Rose, L.S., Akbari, H. and H. Taha, H. (2003). Characterizing the Fabric of the Urban Environment: A Case Study of Greater Houston, Texas. (Paper LBNL-51448.) Lawrence Berkeley National Laboratory, Berkeley, CA.

Santamouris, M. (Ed.). (2001). *Energy and Climate in the Urban Built Environment*. James and James Science Publishers, London, UK.

Santamouris, M. (2013). Using cool pavements as a mitigation strategy to fight urban heat island – A review of the actual developments. *Renewable and Sustainable Energy Reviews*, 26, 224–240.

Santamouris, M. (2014). Cooling the Cities – A Review of Reflective and Green Roof Mitigation Technologies to Fight Heat Island and Improve Comfort in Urban Environments. *Solar Energy*, 103, 682–703.

Santamouris, M., Gaitani, N., Spanou, A., Saliari, M., Gianopoulou, K., Vasilakopoulou, K. (2012a). Using Cool Paving Materials to Improve Microclimate of Urban Areas – Design Realisation and Results of the Flisvos Project. *Building and Environment*, 53, 128–136.

Santamouris, M., Synnefa, A. and Karlessi, T. (2011). Using advanced cool materials in the urban built environment to mitigate heat islands and improve thermal comfort conditions, *Solar Energy*, 85, 3085–3102.

Santamouris. M., Synnefa, A., Kolokotsa, D., Dimitriou, V., Apostolakis, K. (2008). Passive Cooling of the Built Environment – Use of Innovative Reflective Materials to Fight Heat Island and Decrease Cooling Needs. *International Journal Low Carbon Technologies*, 3(2), 71–82.

Santamouris, M, Xirafi, F., Gaitani, N., Spanou, A., Saliari, M., Vassilakopoulou, K. (2012b). Improving the Microclimate in a Dense Urban Area Using Experimental and Theoretical Techniques. – The case of Marousi, Athens. *Int. Journal of Ventilation*, 11(1), 1–16.

Scholz, M. and Grabowiecki, P. (2007). Review of permeable pavement systems. *Building and Environment*, 42, 3830–3836.

Shi, Z. and Zhang, X. (2011). Analyzing the effect of the longwave emissivity and solar reflectance of building envelopes on energy-saving in buildings in various climates. *Solar Energy*, 85(1), 28–37.

Solaimanian, M., Kennedy, T.W. (1993). Predicting maximum pavement surface temperature using maximum air temperature and hourly solar radiation. *Transportation Research Record: Journal of the Transportation Research Board*, 1417, 1–11.

Sthathopoulou, M., Synnefa, A., Cartalis, C., Santamouris, M., Karlessi, T., Akbari, H. (2009). A surface heat island study of Athens using high-resolution satellite imagery and measurements of the optical and thermal properties of commonly used building and paving materials, *J. Sustainable Energy*, 28(1–3), 59–76.

Streutker, D.R. (2002). A remote sensing study of the urban heat island of Houston, Texas. *International Journal of Remote Sensing*, 23, 2595–2608.

Streutker, D.R. (2003). Satellite-measured growth of the urban heat island of Houston, Texas. *Remote Sensing of Environment*, 85, 282–289.

Synnefa, A., Dandou, A., Santamouris, M., Tombrou, M., Soulakellis, N. (2008). Use of cool materials as a Heat Island Mitigation Strategy. *Journal of Applied Meteorology and Climatology*, 47(11), 2846–2856.

Synnefa, A., Karlessi, T., Gaitani, N., Santamouris, M., Assimakopoulos, D.N., Papakatsikas, C. (2011). On the Optical and Thermal Performance of Cool Colored Thin Layer Asphalt Used to Improve Urban Microclimate and Reduce the Energy Consumption of Buildings. *Building and Environment*, 46(1), 38–44.

Synnefa A., Santamouris, M. and Apostolakis, K. (2007). On the development, optical properties and thermal performance of cool colored coatings for the urban environment. *Solar Energy*, 81, 488–497.

Synnefa, A., Santamouris, M. and Livada, I. (2006). A study of the thermal performance of reflective coatings for the urban environment. *Solar Energy*, 80(8), 968–981.

Tadanobu, N. and Tsuyoshi, F. (2010). Cooling effect of water-holding pavements made of new materials on water and heat budgets in urban areas. *Landscape and Urban Planning* 2010, 96, 57–67.

Taha, H., Sailor, D., and Akbari, H. (1992). High albedo materials for reducing cooling energy use. Lawrence Berkeley Lab Rep. 31721, UC-350. Lawrence Berkeley National Laboratory, Berkeley, CA.

Takahashi, T. (2011). Challenge for Cool City Tokyo, Osaka, GSEP Workshop Washington D.C., Sep. 12–13.

Takahashi, K. and Yabuta, K. (2009.) Road temperature mitigation effect of "road cool,"a water-retentive material using blast furnace slag. JFE Technical Report, 13.

Takebayashi, H. and Moriyama, M. (2012). Study on surface heat budget of various pavements for urban heat island mitigation. *Advances in Materials Science and Engineering*, 2012. (Article ID 523051). doi:10.1155/2012/523051

Tetsuji, O. and Suzuki, Y. (1998). Study on the contribution of water-retentive ceramic tile to the reduction of environment heat accumulation. *Journal of Hydraulic Engineering*, 42.

Tran, N. and Powell, B. (2009). Strategies for Design and Construction of High-Reflectance Asphalt Pavements, NCAT Report 09–02.

Tran, N., Powell, B., Marks, H., West, R., Kvasnak, A. (2009). Strategies for Design and Construction of High-Reflectance Asphalt Pavements. *Transportation Research Record: Journal of the Transportation Research Board*, No. 2098, Transportation Research Board of the National Academies, Washington, D.C., 124–130. DOI: 10.3141/2098-13

Wan, W.C., Hien, W.N., Ping, T.P. and Aloysius, A.J.W. (2009). A study on the effectiveness of heat mitigating pavement coatings in Singapore. In: Second International Conference on Countermeasures to Urban Heat Islands, Berkeley, CA, September 21–23.

Yavuzturk, C., Ksaibati, K. and Chiasson, A. D. (2005). Assessment of temperature fluctuations in asphalt pavements due to thermal environmental conditions using a two dimensional, transient finite-difference approach. *J Mater Civ Eng*, 17(4), 465–475.

Yokota, K., Yamaji, T. and Hirano, S. (2010). Basic characteristics of water permeable/retainable porous paving bricks for controlling urban heat island phenomenon. *Journal of Heat Island Institute International*, 5.

Yukari, A. (2009). Development of pervious concrete. (Master's thesis). University of Technology, Sydney Faculty of Engineering and Information Technology.

6

THE EFFECT OF EVAPORATIVE COOLING TECHNIQUES ON REDUCING URBAN HEAT

Servando Alvarez Domínguez[1] and Francisco José Sánchez de la Flor[2]

[1] Universidad de Sevilla – Seville, Spain
[2] Universidad de Cádiz – Cádiz, Spain

Urban climate and water bodies

The influence of water bodies on urban climate is well known, and it has been studied in depth by different authors. On one hand, the absence of water and evapotranspiration from vegetation has been identified as one of the reasons for the so-called 'urban heat island effect' (Oke T. R., 1987). On the other hand, the use of water on the urban environment is one of the most effective ways to remove the urban heat in summer conditions (Santamouris M. and Asimakopoulos D., 1996).

At the same time, water has been studied for its influence on outdoor comfort (Givoni B., 1976; Akbari H. et al., 1992; Santamouris M. *et al.*, 2004), as well as for its impact on the energy consumption of surrounding buildings (Santamouris M. *et al.*, 2001; Sánchez F. J. *et al.*, 2006).

Nowadays, the approach to the problem passes through the empirical results by experimental campaigns, numerical solutions given by detailed software tools and based on CFD simulations, and even the use of satellite images. Thus, the wetlands' impact on the surrounding thermal environment has been analysed, thanks to the use of satellite images. This technique presents an interesting potential to study large areas, and provides practical data to guide decisions in urban landscape design targeted to mitigate UHI effects (Sun and Chen, 2012).

The importance of the presence of water on the UHI effect and, then, its role as a natural cooling technique depends on the climate. This dependency has been identified as relevant for some applications of water to decrease the air temperature, both indoor and outdoor. One of these techniques is based on the use of shower or misting towers when using sprays or micronizers, respectively. The cooling potential of such cooling towers can be expressed in terms of applicability maps (Salmeron J. M. *et al.*, 2012).

Interesting and useful results for urban designers, builders and architects can be obtained from previous experiences and case studies, or by the use of simplified or detailed models (Bruse M. *et al.*, 1998; Sánchez F .J. and Álvarez S., 2004; Asawa T., 2008).

Finally, the interactions between water bodies and other urban cooling techniques have been considered, especially in combination with the evapo-transpiration of vegetation (Robitu M., 2006; Qiu G. *et al.*, 2013; Gago E. J. *et al.*, 2013). A comparison of the cooling potential of different techniques has been also assessed in terms of water use, since this aspect could be crucial in hot and arid climates (Shashua-Bar L. *et al.*, 2009).

Fundamentals of the use of water for urban cooling purposes

Water bodies show several advantages as a natural cooling technique in order to cool the urban environment due to their thermal and optical properties that can be grouped into 3 main effects:

- The high specific heat of water is of an order of about four times the value of most urban materials, and as a consequence, the thermal inertia of water is then four times higher. The effect of the thermal inertia is double: to delay and buffer the maximum temperature.
- The evaporation process of water requires a high amount of energy, known as its latent heat. When a drop of water is evaporated, this energy comes from the surrounding air and water, resulting in this way in a cooler air and water. The minimum reachable temperature for this process is the air wet-bulb temperature.
- The low reflectivity of water causes a low solar reflection to other surfaces in the surroundings, avoiding in this way their warming up.

From all of these effects, the most powerful and then most promising techniques are those based on the evaporative cooling. The amount of energy that is necessary to evaporate water, or latent heat (λ_{water} = 2453 kJ/kg), that in comparison with the energy to heat or cool water, or sensible heat given by its specific heat (sh_{water} = 4.183 kJ/kg K), is almost six hundred higher ($\lambda_{water}/sh_{water}$ = 587 K).

As a matter of example, when the evaporation of water is used to cool the air (sh_{air} = 1.02 kJ/kg K), 1 kg of water can decrease the temperature of more than 2000 m³ of air (ρ_{air} = 1.175 kg/m³) in 1°C, increasing, at the same time, the relative humidity in 5 per cent.

When water is brought into contact with non-saturated hot air, a simultaneous heat and mass transfer process takes place. On one hand, the temperature difference between the air and the water gives rise to a net heat transfer between them. On the other hand, the concentration gradient of water vapour between the water surface and the non-saturated air provokes the evaporation of the water, which requires the use of a certain energy for the phase change. In relation to the thermal level of the water, these two phenomena have an opposed effect. The heat transfer tends to increase the water temperature, whereas the latent energy absorbed by the evaporated water cools the water. The result is an equilibrium temperature called wet-bulb temperature.

Evaporative cooling can be then used for two purposes – to cool the water, or to cool the air:

• **An application of the first is used in fountains**. Cooling by fountains is achieved simultaneously by two different ways, by cooling directly the air of the surrounding space and by cooling the water of the pond. The water pond temperature depends on the existence of sprays, their number and kind, when they operate, if the pond is shaded, and pond depth.

The design process of the spraying system to cool a pond has to consider among others the following aspects (Yannas S. *et al.*, 2006): the range of droplet radius should be of 0.5–1.0 mm; the spray heights in the range 0.5–1.0 m; nozzles should be distributed uniformly around the pond surface, avoiding overlap; and finally, as spraying systems consume water as well as electricity, it is desirable to limit the duration of their operation, taking into account that a spraying system should be turned off when the pond temperature falls below the wet-bulb temperature.

More complex strategies are required when a pond cover is used where considerations related to the operation are: for a movable cover, to open it at night and to close it during daytime; and for a fixed cover, this one should be separated enough from the water level to allow the ventilation needed.

Ponds and fountains can be effective air conditioning systems in open spaces because of their ability to keep water temperatures lower than air temperature, and their low reflectivity. The daily range of water temperature (maximum to minimum difference) is reduced and there is a phase shift between air and water temperatures.

A large set of experiments (Guerra Macho J.J. *et al.*, 1994) were performed in a pond of 30 cm depth, in the Expo '92 grounds (Figure 6.1). Figure 6.2 shows water pond temperature and air temperature in a nearby meteorological station over four consecutive days. The sprayer system worked during the first two days. There were typical ranges of 17°C in air temperature but only 3°C in water temperature with sprayers working and 6°C on the other days. When the sun is shining, the water temperature is always less than the air temperature. On days with the sprayer system working, the water temperature is less than 24°C, while the maximum air temperature is almost 40°C.

Figure 6.1 Fountains at EXPO'92 site in Seville
Source: Guerra Macho J.J. et al., 1992

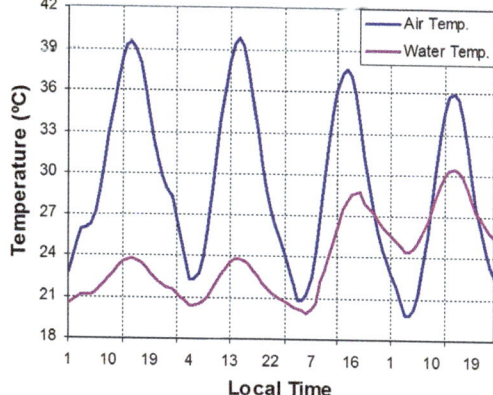

Figure 6.2 Air and water measured temperatures

Figure 6.3 Children's playground at Mairena del Alcor,
Seville
Source: Sánchez F.J. et al., 2010

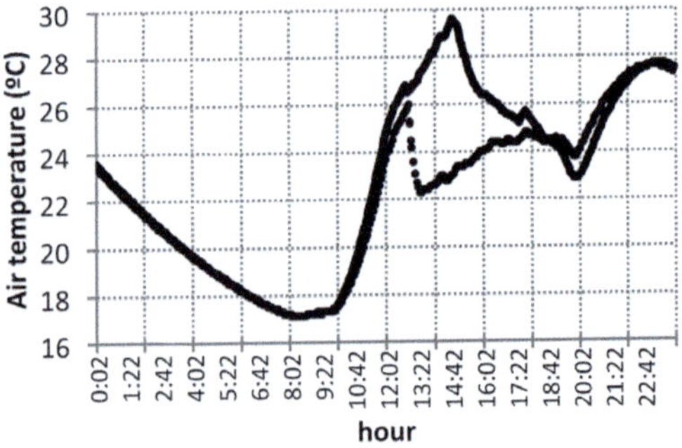

Figure 6.4 Air temperatures at different locations

- **An application of the second one** (to cool the air) is given by the use of sprays or nozzles that can supply water directly into the air at occupant height, or at the top of a tower, inducing a down-draft current of cool air.

An experimental campaign was developed under a national research project focus to show the benefits of using water nozzles at a children's playground in Mairena del Alcor, Seville, Spain (Figure 6.3). During a summer day of 2010, air temperatures at the experimental site (Figure 6.4, dotted line) and at a certain distance (Figure 6.4, solid line) were measured (Sánchez F. J. and Álvarez S., 2010). The following figures show a typical day at which the nozzles were working from 12 p.m. to 8 p.m.

Using cooling towers is a most effective way to cool a higher volume of air, thanks to the confinement provided by the tower itself, and can be increased by the use of fans or wind catchers (Álvarez S. *et al.*, 1991). The use of showers at the top of a tower within a building was also tested experimentally (Givoni B., 1993; Pearlmutter D. et al 1996). In this case, the shower towers are also used to supply the air necessary for ventilation purposes, due to the down-draft air current caused by the air humidification. By the time the air is getting more humidity, its temperature is decreasing, and its density is increasing. As a result, the shower induces an air movement inside the tower from top to bottom, and when this is the only driving force, the tower is called a passive down-draft evaporative cooling tower (PDEC tower).

Twelve of these towers were used along the Avenue I of the Expo site in Seville for the 1992 World Fair, covering a total area of 300 m length and 80 m width. Each tower had a 30 m height and employed high-pressure water misting nozzles (micronizers) to induce downdraught cooling (Figure 6.5). A recent proposal for the outdoor conditioning of the Bahrain Polytechnic University Campus was mainly based on the use of misting towers.

The performance assessment of PDEC towers depends on the climate, the cooling requirements (of the building or outdoor space), and the efficiency of the system itself (Ford B. *et al.*, 2010).

Figure 6.5 Misting towers at Avenue I, Expo'92 site, Seville, Spain
Source: Ford B. et al., 2010

Figure 6.6 Proposal for the Bahrain Polytechnic University Campus

Using a mass of water to improve the urban microclimate at different levels

In the urban context, the effect of water bodies can be examined at two different levels:

- Water as a large-scale urban heat sink: rivers, sea, lakes.
- Water in the street landscaping: ponds with or without fountains.

Water as a large-scale urban heat sink

The Influence of surrounding land or large areas within a city, such as forests, urban parks, sea, lakes or rivers on the surrounding built-up area depends on the nature of the heat sink, but also on the length of the cool area in the direction of the prevailing winds, air velocity and the situation of the area downwind where the cooling effect is examined.

In the case of ponds or rivers, several authors have examined their cooling effect on the surrounding areas. Murakawa *et al.* (1991) measured the temperature distribution near a river and studied the influence on the air temperature of the surrounding areas. It appeared that the influence of the river on the microclimate varied with the width of river and road and the building density in the surrounding area. They also analysed the effects of wind velocity on the drop in air temperature. The air temperature difference between the river and the city area was about 3–5°C on a fine day. The analysed wind velocity was about 1–5 m/s.

Katayama *et al.* (1991) examined the cooling effects of a river and a sea breeze on the thermal environment in a built-up area in summer. The measurements were made along the river and an avenue which went in the direction of the sea-land breeze. The air temperature was lower above the river than above the avenue. The air temperature difference increased almost in proportion to the surface temperature difference. The river was found to be a useful open space to introduce the sea breeze into an urban area.

Sun R. *et al.* (2012) measured the mitigating effects on the urban heat island of Beijing, thanks to the use of satellite ASTER images. They studied 15 wetlands, their shapes and locations, and they found a significant impact on the surrounding thermal environment. Graphic analysis of these images revealed a simple and effective technique to identify temperature gradients, etc.

The same technique was also used to analyse the urban cooling island of 92 parks in Nagoya, Japan by crossing ASTER land surface temperature with information derived from high-spatial-resolution IKONOS satellite data. This study concludes that water bodies have showed insignificant contribution to the production of a cooling island, and in any case, a threshold of at least 2 ha is needed (Cao X. *et al.*, 2010).

The interaction of a river with the heat island of a UK urban area has been studied. Results of the levels of cooling at various locations around the river are presented during spring and summer, showing an average cooling of nearly 1ºC or 2ºC respectively. This cooling effect strongly depends on the urban form in relation to the river. Thus, any benefits of the river are lost quickly where the site is physically cut off from the water by walls or buildings (Hathway E.A. and Sharples S., 2012).

In the POLIS project, computational fluid dynamics (CFD) numerical analysis was performed with results very close to those of the literature mentioned. The results were applied to different situations, such as the influence of the Tagus River (14 km of cooling length) on the Expo '98 grounds in Lisbon (600 m

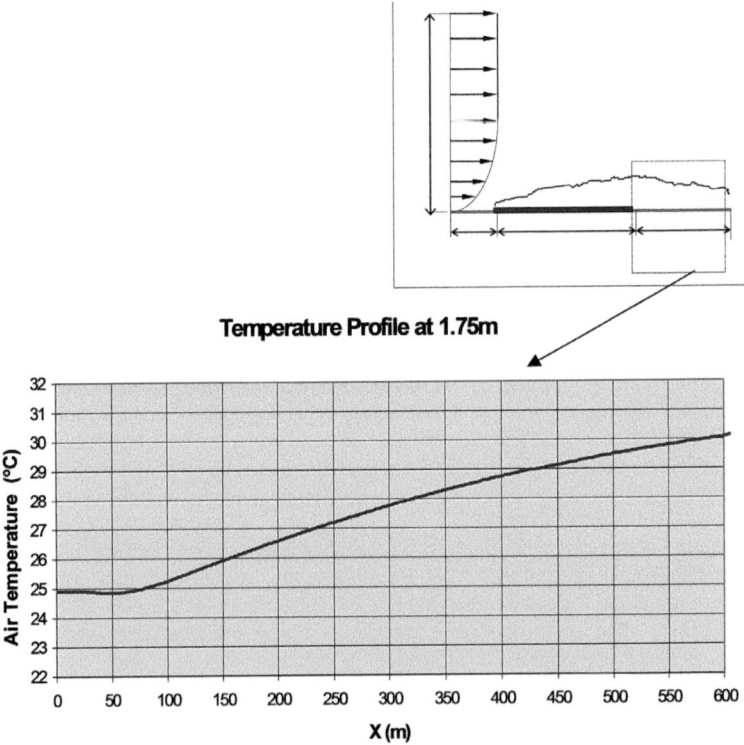

Figure 6.7 **Calculated temperature variation with distance from the River Tagus, Lisbon**
Source: Littlefair P.J., 2000

in the wind direction). Figure 6.7 shows the results for a wind velocity of 1 m/s. There is a temperature drop of about 6°C in the first 100 m due to the presence of the river. From this point, there is a progressive heating of the air so that, at the edge of the site, its temperature is roughly the same as without the cooling effect of the mass of water. For smaller areas of water, however, there is much less impact on the surrounding air temperature (Littlefair P.J. *et al.*, 2000).

Water in street landscaping

At a street scale with street landscaping, the possibilities for the use of mass of waters with the purpose of cooling the ambient air are higher: ponds, fountains, cool towers, etc., can be much more efficient since the cool media is closer to the occupants.

A good example was put into practice in the open areas of the Expo '92 where an extension of 170 hectares of open areas had any kind of cooling treatment based on the use of water, vegetation, covering, or a combination of them (POLIS project, 2000). The Expo '92 took place between April 20th and October 12th 1992.

The objective of these actuations was to get thermal comfort conditions in a city where the maximum air temperature in summer is typically around 40°C. The hottest period covered the months of June to September, with a maximum absolute temperature of 45°C recorded; 1% of the hours in this period (29 hours) were temperatures in excess of 37.8°C, 2.5% exceeded 36.4°C and 5% exceeded 35°C.

The confinement of the zone (protected by the building from the prevailing winds) favours the use of strategies to reduce the air temperature. To get this reduction, a direct evaporative cooling system was implemented based in the use of 128 micronizers under the pergola facing downwards. The flow rate of each micronizer assuming continuous operation was of 7.2 l/h.

These micronizers create an artificial fog by injecting water at high pressure through minute orifices. The small droplets (volume median diameter around 20 mm) evaporate in contact with the surrounding hot air. As the cool air is heavier than the hot air, a continuous descending flow of cool air is obtained.

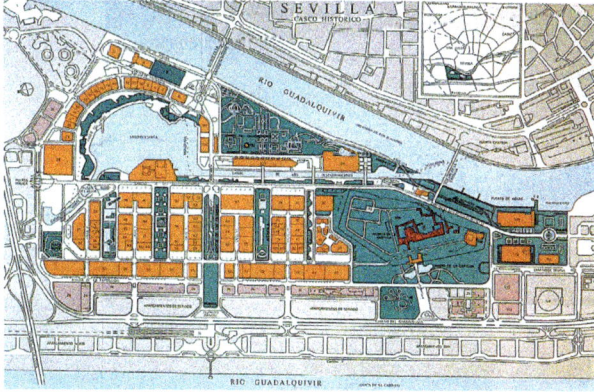

Figure 6.8 EXPO '92 site, Seville, Spain
Source: POLIS project, 2000

Figure 6.9
Air temperature evolution
for three typical summer
days during 1992
Source: POLIS project, 2000

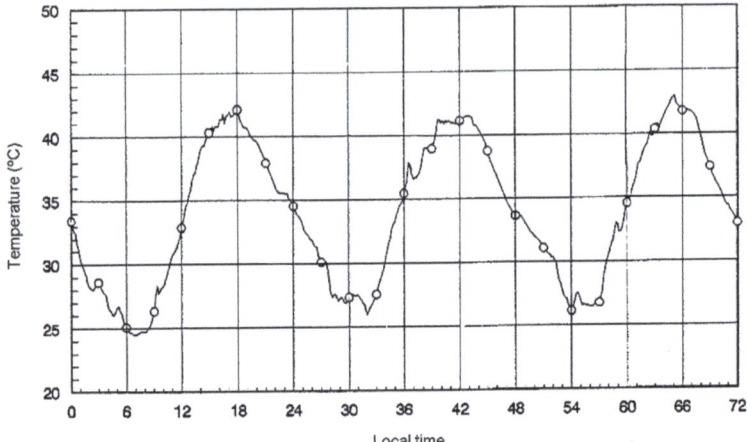

Figure 6.10 Pergola and micronizers
Source: POLIS project, 2000

Figure 6.11 Micronizers
Source: POLIS project, 2000

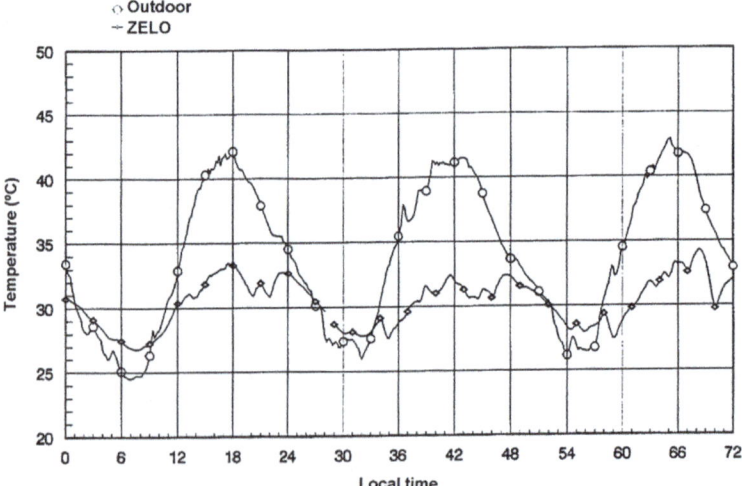

Figure 6.12
Treated and untreated air
temperature
Source: POLIS project, 2000

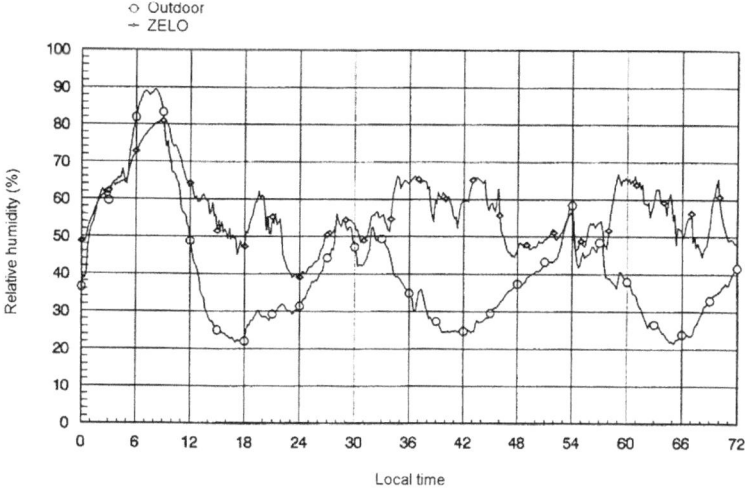

Figure 6.13
Treated and untreated air
humidity
Source: POLIS project, 2000

Figure 6.12 shows a comparison of the outdoor air temperature during 3 summer days in Seville, and in the Expo site during the reference period.

Reductions of about 5°C were found at the peak time hour.

The micronizers operated following an intermittent strategy for two reasons:

1. To avoid getting people wet because of non evaporated water droplets.
2. To maintain the relative humidity of the zone below the level of 65%.

Measures taken during the whole exhibition period revealed that the air temperature was always kept under outdoor air temperature. The temperature of the treated zone air experienced maximal decreasing up to 10°C, from a maximal value of outdoor air of 42°C.

The relative humidity was maintained under 65%, by using the above quoted strategy of use of the micronizers.

The comfort achieved was measured during the whole period revealing that the sweating ratio in the Expo '92 site was kept less than 90 g/h. This assessment was carried out by the use of Fanger's comfort model (Fanger P.O., 1972).

Mathematical models

Energy balance of a water pond

The water temperature and cooling potential of a water pond are the unknowns that have to be solved from the energy balance of a water pond. This energy balance can be expressed for top and bottom surfaces as well as for the water mass, or can include the whole volume and surfaces in a single equation.

The energy balance at the fountain bottom takes into account the energy exchange with the ground, the shortwave radiation impinging the fountain bottom, and the convective exchange between the water and the ground surface.

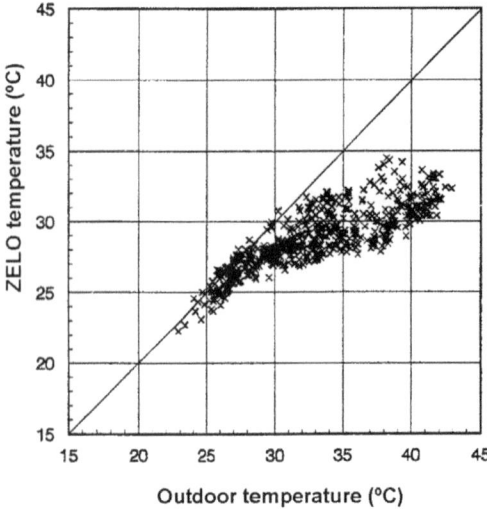

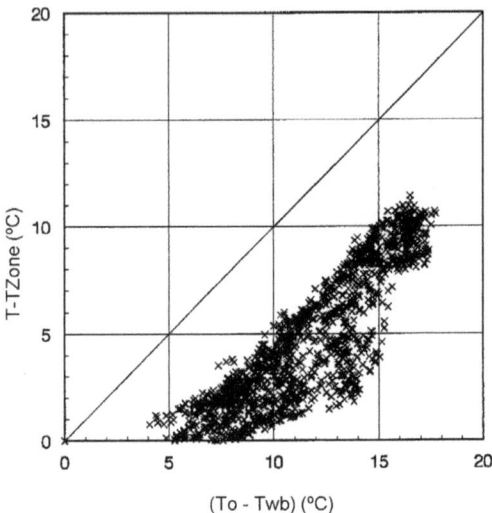

Figure 6.14 Air temperatures comparison
Source: POLIS project, 2000

Figure 6.15 Air temperature drop versus the maximum achievable
Source: POLIS project, 2000

$$Q_{conduction} = R_{ShortWave}^{Bottom} + A \cdot h_{water} \left(T_{Water}^{Bottom} - T_{Surface}^{Ground} \right)$$

At the fountain top, the energy balance considers the convective exchange between the water and the air, the longwave radiation exchange with the sky, the evaporation heat losses from the pond surface, and from the sprays. The last one is the most relevant heat flux in order to cool the fountain.

$$Q_{PondSurface}^{Convection} + R_{LongWave}^{Top} + Q_{PondSurface}^{Evaporation} + Q_{Sprays} = 0$$

Finally, in the whole volume of the fountain, the energy balance includes the heat flux due to the water reposition, and to the absorbed shortwave radiation.

$$R_{ShortWave}^{Absorbed} + Q_{Reposition}^{Water} = 0$$

The solution of these 3 equations allows the calculation of 3 different unknowns, which typically are the water temperature at the top and the bottom, and the water temperature representative of the rest of the fountain.

A more detailed model should consider different water temperatures depending on the fountain depth.

A dynamic model was developed in the context of the ROOFSOL Project (Molina J.L. and Rodríguez E.A., 1998) to study the thermal performance of a shallow water pond, and where the entire water mass was assumed to have the same temperature. This pond temperature was then a unique function of time.

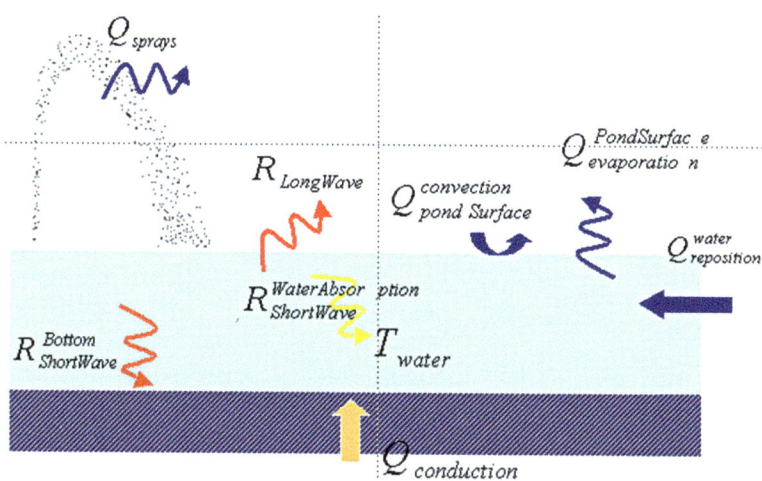

Figure 6.16
Energy exchange in a
fountain

The dynamic thermal performance gives a thermal storage that can be assessed from:

$$\Delta U = \frac{\rho \cdot c_p \cdot V}{A} \cdot \frac{dT}{dt}$$

Where ΔU is the heat stored in the pond per unit pond surface area and time, and can be obtained by the energy balance of the whole control volume as:

$$\Delta U = \left(R_{ShortWave}^{Bottom} + Q_{Reposition}^{Water} \right) - \left(R_{LongWave}^{Top} + Q_{PondSurface}^{Evaporation} + Q_{PondSurface}^{Convection} + Q_{Sprays} + Q_{conduction} \right)$$

Where the main assumptions are:

• **For the attenuation of incident solar radiation in the water pond**, the air-water interface can be considered flat so that Fresnel's equations for reflectance are applicable. This interface reflects specularly, while the bottom surface reflects diffusely.

Solar radiation absorption depends on solar incident angle (α) and the pond depth (H), while solar reflection depends only on the solar incident angle. Figures 6.17 and 6.18 show these dependencies and how both values remain constant for a wide range of values from 0° (normal incidence) to 60°. The absorption coefficient quickly increases with pond depth for shallow water pond but it is stabilised for a pond depth over 1 m.

$$R_{ShortWave}^{Bottom} = \alpha_{water\,mass} \cdot \left(IDH + IdH \right)$$

Where

IDH   is the incident direct radiation over an horizontal surface
IdH   is the incident diffuse radiation over an horizontal surface

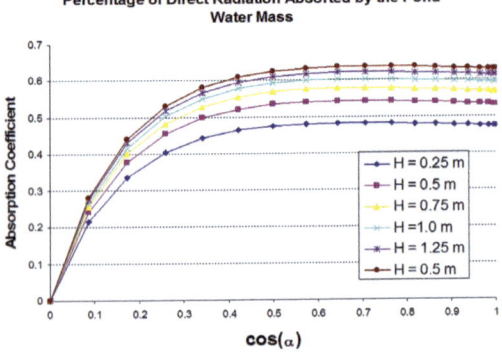

Figure 6.17 Absorption coefficient of the pond water mass   Figure 6.18 Reflectivity factor of the pond surface

• **For the longwave radiation from the top surface,** an emissivity of 0.95 was taken. The sky can be considered as a black body at an effective temperature of 10ºC under air temperature. The last one can be better approached by using more complex models (Berdahl P. and Fromberg R., 1982).

$$R^{Top}_{LongWave} = \sigma \cdot \varepsilon \cdot \left(T^4_{water} - T^4_{sky} \right)$$

Where

σ is the Stefan-Boltzmann constant $\sigma = 5.67e - 8\ W/m^2\ K^4$

• In general, **the convective heat transfer is due to forced convection,** while at low wind speeds, free convection prevails. The first one depends on the air velocity while the second one depends on the difference of temperature between the air and the water. As a result, higher values of convective heat transfer can be expected for wind driven forces.

$$Q^{Convection}_{PondSurface} = h_c \cdot \left(T_{water} - T_{air} \right)$$

Where

h_c is the heat transfer coefficient

• **The amount of heat lost due to water being evaporated from the pond top surface** is given in terms of the mass of water evaporated and the latent heat. As for the convective heat transfer, higher values can be expected for a high wind speed.

$$Q^{Evaporation}_{PondSurface} = \frac{K_m}{R \cdot T_{air}} \cdot h_{lg} \cdot \left(P_{sat}(T_{water}) - RH_{air} \cdot P_{sat}(T_{air}) \right)$$

Where

K_m is the mass transfer coefficient calculated using empirical correlations
R is the universal ideal gas constant

h_{lg} is the latent heat of vaporisation of water
$P_{sat}(T_{water})$ is the water saturation pressure at water pond temperature
$P_{sat}(T_{air})$ is the water saturation pressure at air temperature
RH_{air} is the air relative humidity

• **Heat loss from operation of water sprays** can be estimated from a separate model for an isolated drop which evaporates in a mass of air. Using such a model, it is possible to obtain the final drop radius and the final drop temperature. As the initial drop radius and temperature have to be input data for this model, the total amount of evaporated water and the heat loss from water sprays can be calculated. This model is described below.

• The previous results of water evaporation from the top surface and the operation of water sprays imply **the necessity of an addition of water**. This addition of water is also a heat input in the energy balance that will be typically positive, and depends on the temperature of the water supply.

$$Q_{Reposition}^{Water} = m_{water}^{Evaporated} \cdot c_{p,water} \cdot \left(T_{water}^{supply} - T_{water} \right)$$

• Finally, the presence of a pond cover modifies most of the energy fluxes:

The incident solar radiation is affected by the cover shortwave transmittance. If exists, the pond cover shortwave transmittance should be as low as possible, avoiding or minimising this heat entrance.

The longwave radiation from the top surface to the sky is in this case much lower, since with a pond cover the actual longwave radiation exchange is between the water pond surface and the bottom pond cover surface, and it is the upper pond cover surface which has a direct exchange with sky. For this reason, movable pond covers should be considered and then the operation strategy consists of putting it on during daytime and removing it at night.

A second energy balance has to be imposed at the covering, where a unique temperature, without heat storage, is considered.

Both convective heat flux and evaporation from the top surface are reduced, with a pond cover, due to the expected lower wind speeds.

Energy balance of a water drop

As the partial evaporation of the water drop is mainly responsible for the cooling effect of a water pond, detail mass and energy models have to be used.

The objective of these models is then to calculate the final temperature of the water drop (T_f) and the final radius (T_r).

Conservation equations of mass, energy and momentum are used with the following assumptions:

• The water drop is considered to be a sphere.
• There are not interactions among different water drops.
• The whole volume of the water drop is at the same temperature.

- Air temperature and pressure remain constant for each water drop.
- Negligible radiant heat fluxes.

$$\rho_{water} \cdot \frac{dr}{dt} = \rho_{air} \cdot K_m \cdot \frac{Y_{Ao} - Y_{A\infty}}{1 - Y_{Ao}}$$

$$\frac{dv_x}{dt} = \frac{3}{8} \cdot f \cdot \frac{v \cdot v_x}{r} \cdot \frac{\rho_{air}}{\rho_{water}}$$

$$\frac{dv_y}{dt} = g \cdot \left(\frac{\rho_{air}}{\rho_{water}} - 1 \right) - \frac{3}{8} \cdot f \cdot \frac{v \cdot v_y}{r} \cdot \frac{\rho_{air}}{\rho_{water}}$$

$$\rho_{water} \cdot c_{p,water} \cdot \frac{dT}{dt} = 3 \cdot \frac{h}{r} \cdot \left(T_{air} - T \right) - \frac{3 \cdot \rho_{air} \cdot K_m}{r} \cdot h_{lg} \cdot \frac{Y_{Ao} - Y_{A\infty}}{1 - Y_{A\infty}}$$

Where

r is the drop radius
T is the water drop temperature
Y_{Ao} is the water mass fraction at the drop surface
$T_{A\infty}$ is the water mass fraction at the air
v_x, v_y are the drop velocities in x and y directions

To solve the previous equations set, initial drop radius (r_i) and drop temperature (T_i) should be known, and final drop radius (r_f) and drop temperature (T_f) are obtained. This final temperature of the water drop has a minimum possible value equal to the wet-bulb temperature of the air ($T_f > T_{air}^{wet-bulb}$).

Evaporated volume of water (V_e) and heat loss (Q_e) of a single water drop can be obtained from:

$$V_e = \frac{4}{3} \cdot \pi \cdot \left(r_i^3 - r_f^3 \right)$$

$$Q_e = \frac{4}{3} \cdot \pi \cdot \rho_{water} \cdot c_{p,water} \cdot \left(r_i^3 \cdot T_i - r_f^3 \cdot T_f \right)$$

Then, for a number of water drops (N_d) given by the mass flow circulating through the sprayers (m'):

$$N_d = \frac{m'}{\frac{4}{3} \cdot \pi \cdot r_i^3}$$

Finally, heat loss from operation of water sprays can be estimated from the initial and final drop radius and the initial and final drop temperature. The total amount of evaporated water and the heat loss from water sprays can be calculated as follows:

$$Q_{sprays} = m' \cdot \rho_{water} \cdot c_{p,water} \cdot \left(T_i - \frac{r_f^3}{r_i^3} \cdot T_f \right)$$

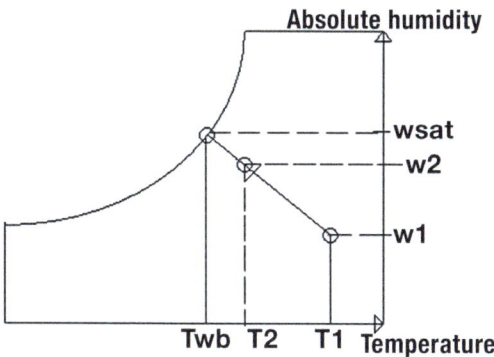

Figure 6.19 Air temperature evolution

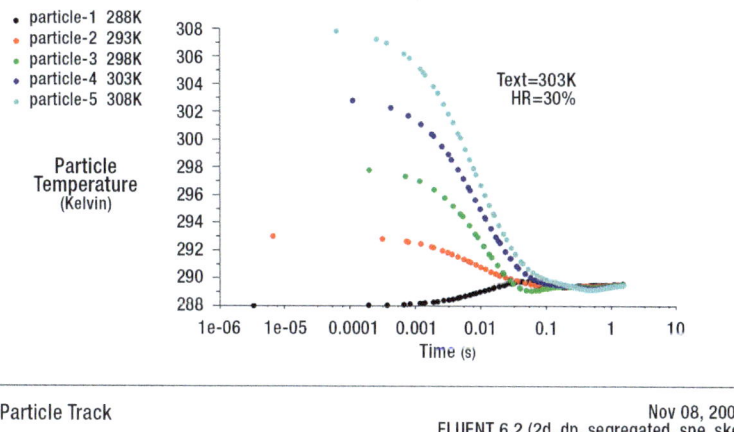

Figure 6.20 CFD simulations of the water drop evolution in a spray

$$m_{sprays}^{Evaporated} = m' \cdot \left(1 - \frac{r_f^3}{r_i^3} \right)$$

Figure 6.19 shows the air temperature evolution in contact with a mass of water, and Figure 6.20 shows the water drop temperature as a function of the initial drop temperature and time. As can be seen, the final drop temperature is independent from the initial conditions, since wet-bulb temperature is reached in all cases.

Potential of evaporative cooling to reduce the temperature of the air

The evaporative cooling potential mainly depends on the dry (DBT) to wet (WBT) air temperature depression. In terms of this difference, applicability maps of water evaporation as a natural cooling technique can be drawn.

Using synthetic climatic data obtained from the software Meteonorm for known locations, and an extrapolation technique for the rest (Sánchez *et al.*, 2008), necessary data can be obtained, and then processed to assess applicability indexes like:

- Average dry to wet air temperature. This value (in ºC) should be calculated for the summer period, when the evaporative cooling technique can be applied.

$$\overline{DBT - DWT} = \frac{1}{N_{hours}} \sum_{i=1}^{N_{hours}} \left(T_{air}^{DBT}(i) - T_{air}^{WBT}(i) \right)$$

- A simple modification of the previous leads to the definition of cooling degree-days.

$$CDD(DBT - DWT) = \frac{1}{24} \sum_{i=1}^{N_{hours}} \left(T_{air}^{DBT}(i) - T_{air}^{WBT}(i) \right)$$

- Frequency in hours of the DBT-WBT temperature depression during the summer season.

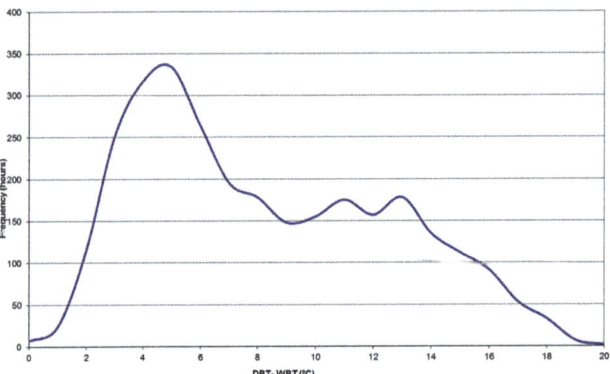

Figure 6.21
Frequency in hours of the DBT-WBT temperature depression for Seville from June to September

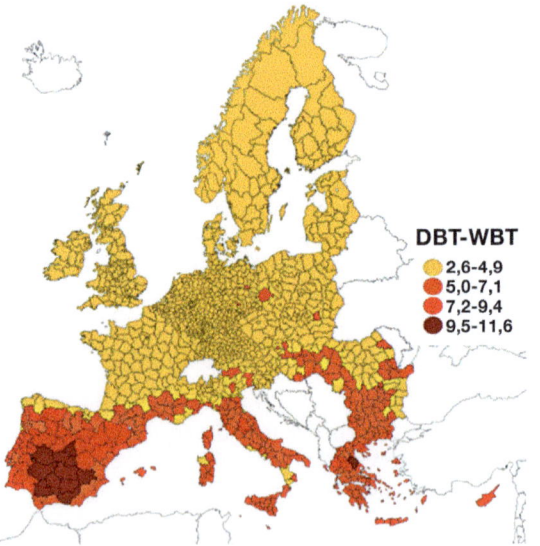

Figure 6.22
Applicability map of the evaporative cooling technique in Europe
Source: Salmerón J.M. et al., 2012

References

Akbari H., Davis S., Dorsano S., Huang J. and Winert S. (1992). Cooling our Communities – A Guidebook on Tree Planting and Light Colored Surfacing. US Environmental Protection Agency, Office of Policy Analysis, Climate Change Division.

Álvarez S., López de Asiaín J., Yannas S. and Oliveira E. (1991, 24–27 Sept.). Architecture and Urban Space. Proc. 9th PLEA Int. Conf., Seville, Spain.

Asawa T., Hoyano A. and Nakaohkubo K. (2008.) Thermal design tool for outdoor spaces based on heat balance simulation using a 3D-CAD system. *Building and Environment*, 43, 2112–2123.

Berdahl P. and Fromberg R. (1982). The thermal radiance of clear skies. *Solar Energy*, 29, 4, 299–314.

Bruse M., Fleer H. (1998). Simulating surface-plant-air interactions inside urban environments with a three dimensional numerical model. *Environmental Modelling & Software*, 13, 373–384.

Cao X., Onishi A., Chen J. and Imura H. (2010). Quantifying the cool island intensity of urban parks using ASTER and IKONOS data. *Landscape and Urban Planning*, 96, 224–231.

Fanger P.O. (1972). *Thermal Comfort*. McGraw-Hill Book Company, New York.

FLUENT(r) Academic Research Computational Fluid Dynamics Software Tool, Release 6.2. (2005).

Ford B., Francis E., Álvarez S., Thomas P., and Schiano-Phan R. (2010). *The Architecture and Engineering of Downdraught Cooling: A Design Sourcebook*. PHDC Press.

Gago E.J., Roldan J., Pacheco R. and Ordóñez J. (2013). The city and urban heat islands: A review of strategies to mitigate adverse effects. *Renewable and Sustainable Energy Reviews*, 25, 749–758.

Givoni B. (1976). *Man, Climate and Architecture*. Elsevier, New York.

Givoni B. (1993). Semi-empirical model of a building with a passive evaporative cooling tower. *Solar Energy*, 50, 425–434.

Guerra Macho J.J., Cejudo Lopez J.M., Molina Felix J.L., Alvarez Domínguez S. and Velázquez Vila. (1994). Control Climático en Espacios Abiertos: Evaluación del proyecto Expo '92. Sevilla: Universidad de Sevilla, Ed. Publicaciones CIEMAT.

Hathway E.A. and Sharples S. (2012). The interaction of rivers and urban form in mitigating the Urban Heat Island effect: A UK case study. *Building and Environment*, 58, 14–22.

Katayama T., Ishii A., Nishida M., Tsutsumi J. and Oguro M. (1991). Cooling Effects of a River and Sea Breeze on the Thermal Environment in a Built-up Area. *Energy and Buildings*, 15–16, 973–978.

Littlefair P.J., Santamouris M., Álvarez S., Dupagne A., Hall D., Teller J., Coronel J.F. and Papanikolaou N. (2000). *Environmental site layout planning: solar access, microclimate and passive cooling in urban areas*. BRE Publications, London, U.K.

Meteonorm, Version 7.1.5. Global meteorological database for solar energy and applied climatology. Bern, Switzerland: Meteotest.

Molina J.L. and Rodríguez E.A. (1998). Detailed modelling of water ponds. In Final Research Report of Task 2: Modelling Cooling Roofs and Design Tool Development. ROOFSOL Project European Commission Joule Programme JOR3CT960074.

Murakawa S., Sekine T. and Narita K. (1991). Study of the Effects of a River on the Thermal Environment in an Urban Area. *Energy and Buildings*, 15–16, 993–1001.

Oke T.R. (1987). *Boundary Layer Climates*. Routledge, Cambridge, U.K.

Pearlmutter D., Erell E., Etzion Y., Meir I.A. and Di H. (1996). Refining the use of evaporation in an experimental down-draft cool tower. *Energy and Buildings*, 23, 191–197.

POLIS project final report. (2000). European Commission Directorate General XII Science, Research and Development.

Qiu G., Li H., Zhang Q., Chen W., Liang X. and Li X. (2013). Effects of evapotranspiration on mitigation of urban temperature by vegetation and urban agriculture. *Journal of Integrative Agriculture*, 12, 8, 1307–1315.

Robitu M., Musy M., Inard C. and Groleau D. (2006). Modeling the influence of vegetation and water pond on urban microclimate. *Solar Energy*, 80, 435–447.

Salmerón J.M., Sánchez F.J., Sánchez J., Álvarez S., Molina J.L. and Salmerón R. (2012). Climatic applicability of downdraught cooling in Europe. *Architectural Science Review*, 2012, 1–14.

Sánchez F.J. and Álvarez S. (2004). Modelling microclimate in urban environments and assessing its influence on the performance of surrounding buildings. *Energy and Buildings*, 36, 403–413.

Sánchez F.J. and Álvarez S. (2010). CAVIARU national research project interim report. Experimental campaign at Mairena del Alcor, Seville, 2010 summer.

Sánchez F.J., Álvarez S., Molina J.L. and González R. (2008). Climatic zoning and its application to Spanish building energy performance regulations. *Energy and Buildings*, 40, 1984–1990.

Sánchez F.J., Salmerón J.M. and Álvarez S. (2006). A new methodology towards determining building performance under modified outdoor conditions. *Building and Environment*, 41, 1231–1238.

Santamouris M., Adnot J., Álvarez S., Klitsikas N., Orphelin M., Lopes C. and Sánchez F. (2004). *Cooling the cities – Rafraichir les villes*. Les Presses des Mines, Ecole des Mines, Paris.

Santamouris M. and Asimakopoulos D., (Eds.) (1996). *Passive Cooling of Buildings*. James and James Science Publishers, London, U.K.

Santamouris M., Papanikolaou N., Livada I., Koronakis I., Georgakis C., Argirou A., Assimakopoulos D.N. (2001). On the impact of urban climate on the energy consumption of buildings. *Solar Energy*, 7, 3, 201–216.

Shashua-Bar L., Pearlmutter D. and Erell E. (2009). The cooling efficiency of urban landscape strategies in a hot dry climate. *Landscape and Urban Planning*, 92, 179–186.

Sun R., Chen A., Chen L. and Ju Y. (2012). Cooling effects of wetlands in an urban region: The case of Beijing. *Ecological Indicators*, 20, 57–64.

Sun R. and Chen L. (2012). How can urban water bodies be designed for climate adaptation? *Landscape and Urban Planning*, 105, 27–33.

Yannas S., Erell E. and Molina J.L. (2006). *Roof Cooling Techniques: A Design Handbook*. Earthscan.

7

EXPLOITING EARTH COOLING TO MITIGATE HEAT ON CITIES' SCALE

Stamatis Zoras and Argiro Dimoudi

Democritus University of Thrace – Komotini, Greece

Preface

Supply of cool air during the summer period in open areas is utilised to improve human thermal comfort by the urban microclimate improvement of populated spaces. The aim of the chapter is to study the implementation of earth cooling techniques, in open spaces, that mitigate heat in a city's scale. Details of the earth's thermal inertia in contact with the atmosphere will be analysed together with the basic underground technologies, e.g. underground heat exchange tubes for external environment improvement. Boundary conditions, the basic equations that describe the heat transfer and air movement and simulation tools that handle microclimatic effects due to earth cooling will also be discussed. Practical information related to sizing and construction of these systems will be tackled. The chapter concludes by presenting examples of application of these systems at outdoor spaces.

Introduction

Significant energy is stored in the ground throughout the year, which could be used in the cooling of ambient air. It is known that ground temperature changes more slowly than ambient air's temperature and it becomes stable with depth. Due to the high thermal inertia of the soil, temperature fluctuations at the ground surface are attenuated inside the ground and a time lag is observed between the surface and the ground temperature. The greater the depth in the ground, the more attenuation of the ambient temperature is observed and at a certain depth, ground temperature is lower than the ambient temperature in summer and is higher in winter (Pfafferott *et al.*, 2007).

The principle of using soil thermal inertia dates back to the ancient Greeks and Persians (Argiriou, 1996) with underground building structures or earth tubes (Delos earth pipes network). In Medieval Italy, special natural caves called 'covoli' were used to pre-cool or preheat air before it entered a building (Argiriou, 1996). Implementation of earth tubes in building structures has become fairly common since the mid-1990s in Austria, Denmark, Germany, and India and is slowly being adopted in North America. In the very last years, earth tubes were proposed for bioclimatic redevelopment of outdoor spaces (Gaitani *et al.*, 2011; Fintikakis *et al.*, 2011; Santamouris *et al.*, 2012) in order to mitigate the heat island effect.

Earth coupled systems increase the cooling rate of outdoor spaces via convective cooled air. Specifically, the air from ground heat exchangers cools ambient environment via turbulent fluid mixing. Convective heat diffusion is a process that affects air temperature and may be employed in the mitigation of urban heat islands.

The earth to air domain

Air is cooled by its circulation at soil's depth, i.e. 3 m (Mihalakakou et al., 1992), within ground heat exchangers and then is diffused into ambient air, reducing this way the ambient temperature. The relatively low temperature of soil cools down the air in underground tubes. Many different tools have been developed for the prediction of heat transfer in earth-coupled structures and buried tubes. Each one of them has its own advantages and limitations. A recent review (Zoras, 2009) reports on the methods and tools that cope with the calculation of heat transfer to the ground from structures on contact with the soil.

Earth-contact domain

Heat conduction governs the solid domain that depends on soil's thermal properties. Air in buried tubes is cooled due to earth contact from cooler soil during summer. The operation and efficiency of ground heat exchangers have been widely studied by Mihalakakou et al., (1994a, b). Installed diffusers deliver cool air from underground tubes into ambient air.

A variety of modeling and experimental techniques of the soil's temperature calculation have been developed in the past, which mostly require past field data for validation and model construction (Mihalakakou, 2002) while based on a one-dimensional transient heat conduction equation (Mihalakakou et al., 1997). However, this method of soil temperature calculation seems limited due to its one-dimensional nature (Hollmuller and Lachal, 2014). With regard to the earth-contact component, it is necessary to recognise that the soil energy exchange process is complex, i.e. multi-dimensional (Davies et al., 1995). Many alternative simulation methods have been proposed which deal with this complexity with varying degrees of sophistication (for an extensive review see Adjali et al., 1998a).

The complexity of earth-coupled systems inevitably leads to assumptions being made regarding the parameters that influence the processes of heat transfer to the ground. However, implications arise due to the non-linear nature of heat transfer concerning moisture transfer and thermal conductivity in relation to temperature variation and other climatic conditions, e.g. snow and rain (Adjali et al., 1998b).

Thermal properties of soils

The heat capacity of a material is required when non-steady solutions are to be determined. In effect, the heat capacity, C_p, defines the amount of energy stored in a material per unit mass per unit change in temperature ($J\,kg^{-1}\,K^{-1}$). It is given as a function dependent on the different heat capacities and volumes of the constituents (Rees et al., 2000):

$$C_p = v_1 \rho_1 C_{P_1} + v_2 \rho_2 C_{P_2} + \ldots \tag{1}$$

where C_p is the total heat capacity of the material and v (m^3), ρ ($kg\,m^{-3}$) and C are, respectively, the volume fractions, the densities and the heat capacities for each constituent.

The thermal conductivity of a material, λ, is the constant of proportionality that relates the rate at which heat is transferred by conduction to the temperature gradient inside the material (W m^{-1} K^{-1}). In fact, to simulate an earth-coupled system, a problem arises regarding the difficulty in defining the thermal conductivity of the soil. The thermal conductivity of the soil depends on temperature and the different materials (soil grains, water and air) that compose the ground. The soil moisture content influences the temperature of the ground by its transportation and by the latent heat caused from phase changes. Saturated soils transmit heat at a faster rate than unsaturated materials. A good assumption is to neglect the air's thermal conductivity since it is much lower than the conductivities of the other components. It is extremely difficult to simulate the ground to take into account both moisture and temperature changes. So, usually a constant ground temperature and a fixed amount of moisture are utilised, which results in inaccuracies in cases where soil property combinations are of paramount importance. Thus, the most efficient tool would include the calculation of coupled heat and moisture transfer in the ground. Theoretically, for an isotropic material, the thermal conductivity can be expressed as:

$$\lambda = \frac{Qd}{\left(T_o - T_1\right)St} \tag{2}$$

where λ (W m^{-1} K^{-1}) is the thermal conductivity, d is the thickness of a plate in the soil (m), Q the heat (J) which flows up through the plate in t seconds from a surface S (m^2) caused by a difference in temperatures T_o–T_1 (K), where T_o and T_1 are the temperatures of the two sides of the defined solid plate (Carslaw and Jaeger, 1959). This temperature gradient applies after the soil has reached its steady state with no further significant changes in temperature values. Generally, the calculation of thermal conductivity of soils is not a straightforward and easy procedure due to the generally complex nature of soils (e.g. anisotropy, soils composed of more than one material). However, the definition of conductivity is always restricted by assumptions being made of homogeneous, isotropic materials and moisture content (Rees et al., 2000).

The thermal resistance R (m^2 K W^{-1}) of a structural component is the property that characterises the response of the component to heat transfer fluctuations. For a unit area of a slab of homogeneous material, the thermal resistance is calculated by dividing its thickness, l (m), by its thermal conductivity, λ (W m^{-1} K^{-1}) (CIBSE, 1980), and it Is expressed as:

$$R = \frac{l}{\lambda} \tag{3}$$

However, the thermal resistance of a composite solid can be determined by adding the individual resistances of different materials. The heat transfer by radiation and convection at the outer or inner surfaces of building elements or the ground surface can be treated as flow through a thermal resistance R_S (m^2 K W^{-1}) given by the following equation (CIBSE, 1980):

$$R_S = \frac{1}{Eh_r + h_c} \tag{4}$$

where E is the emissivity factor, h_r (W m^{-2} K^{-1}) the radiative heat transfer coefficient and h_c (W m^{-2} K^{-1}) the convective heat transfer coefficient.

The thermal transmittance (U-value) of a structural element gives an indication of the mean rate of heat transfer through the component. It is defined by the reciprocal of the summation of the individual thermal resistances that compose the building element. Therefore, the U-value (W m^{-2} K^{-1}) is given by the following equation (CIBSE, 1980):

$$U = \frac{1}{R_{SI} + R_1 + R_2 + \ldots + R_A + R_{SO}} \qquad (5)$$

where R_{SI} (m^2 K W^{-1}) is the inside surface resistance, R_1, R_2 . . . (m^2 K W^{-1}) the thermal resistances of structural components, R_A (m^2 K W^{-1}) the airspace resistance and R_{SO} (m^2 K W^{-1}) the outside surface resistance.

Numerical heat transfer

The categories of tools in the estimation earth-contact transfer are (Zoras, 2009):

a)   analytical and semi-analytical methods
b)   numerical methods
c)   manual methods and design guides.

Generally, analytical solutions under certain conditions, can give accurate and fast results for earth-contact heat transfer. The drawback of these solutions is apparent when the soil properties are not constant in space and time, the geometry is complex and the entire phenomena taking place in the procedure of heat transfer must be included. The manual methods have limited applications because of their simplified nature (e.g. reduction of the number of parameters which influence the phenomena or assuming some parameters constant). However, in many cases, these methods can yield general conclusions regarding the thermal performance of some buildings.

Numerical methods deal with the solution of differential equations (e.g. heat conduction equation) at discrete points that define the grid in which the domain is depicted. To solve the equations at each grid point the use of a general method is necessary (e.g. Taylor series, weighted residuals). The way that the phenomena are simulated according to conservation (of energy), space area definition (geometry) and continuity (in space and time) depends on the method or the combination of methods being used. Numerical simulation seems the most promising way to handle multi-dimensional earth-contact heat transfer.

Heat transfer takes place via:

1.   Conduction, in which activity at molecular level can generate a flow of heat through any kind of material (e.g. solid, liquid, vapour or gas).
2.   Convection, in which heat is transferred by relative motion of portions of the heated body.
3.   Radiation, in which heat is transferred directly between distant portions of the body by electromagnetic radiation.

In liquids and gases, convection and radiation are of paramount importance, but in solids, convection is altogether absent and radiation is usually neglected (Carslaw and Jaeger, 1959).

Figure 7.1 describes the air circulation in the conductive soil and the governed phenomena at each phase. Conduction is the prevailing process in the soil that

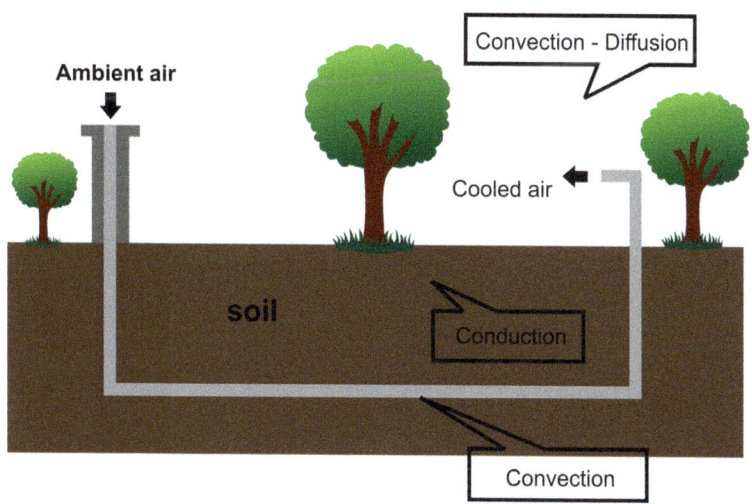

Figure 7.1
Ground heat exchanger
installation for an open
space

drives air temperature reduction in the tube. Airflow in the tube is described by convection, while in the ambient air, convection and diffusion must be taken into consideration. Earth surface temperature continuously affects ambient air via convection (air to heat exchanger's surface) and radiation (between opposite sides of heat exchanger's surfaces). In the following theoretical analysis, the mathematical approximation of the described earth-to-air heat exchange system is presented.

Taylor series formulation

Taylor series provides a means to predict a function value at one point in terms of the function value and its derivatives at another point. It is based on the Taylor series theorem, which says that any smooth function can be approximated as a polynomial. If a function $f(x)$ is known at the point x_i, then it can be approximated at a point x_{i+1} by adding the derivatives of the function at point x_i in a series expansion. The first term added represents a zero-order approximation and indicates that the value of the function at the new point is the same as its value at the old point:

$$f(x_{i+1}) = f(x_i) \tag{6}$$

The zero-order approximation is a good one for constant functions. Actually, equation (6) could be a good approximation if the two points are very close to each other. The second term that can be added in the above equation depicts a first-order approximation, and it is the slope of the function at point x_i. This is multiplied by the distance between the points in order to give an increase or decrease of the function:

$$f(x_{i+1}) = f(x_i) + f'(x_i)(x_{i+1} - x_i) \tag{7}$$

The second-order approximation term reveals the curvature that the function might exhibit:

$$f(x_{i+1}) = f(x_i) + f'(x_i)(x_{i+1} - x_i) + \frac{f''(x_i)}{2!}(x_{i+1} - x_i)^2 \tag{8}$$

In the same manner, additional terms are added so that the approximation becomes more accurate:

$$f(x_{i+1}) = f(x_i) + f'(x_i)(x_{i+1} - x_i) + \frac{f''(x_i)}{2!}(x_{i+1} - x_i)^2 + \dots \qquad (9)$$
$$+ \frac{f^{(n)}(x_i)}{n!}(x_{i+1} - x_i)^n + R_n$$

The R_n term includes terms from $n+1$ to infinity. However, the definition of higher order derivatives is difficult and the approximation is generally restricted to a finite number of terms. The remaining terms are included in the R_n term called *truncation error*. This term represents the discrepancy introduced by the fact that numerical models employ approximations to represent exact mathematical operations and quantities, and it is given by the equation:

$$R_n = \frac{f^{(n+1)}(\xi)}{(n+1)!}(x_{i+1} - x_i)^{n+1} \qquad (10)$$

where ξ is a value of x that lies between x_i and x_{i+1}. As shown in equation (10), the truncation error is of the order of $(x_i - x_{i+1})^{n+1}$ and thus, the error is proportional to the step size used to approximate the function at the new point. From this, it can be concluded that if the error is of the order of $(x_i - x_{i+1})$, then halving the step size will halve the error. If the error is of the order of $(x_i - x_{i+1})^2$, then halving the size step will quarter the error.

The finite difference method
This method, at each grid point, approximates the partial derivatives of the variables by replacing them with approximations in terms of the nodal values of the functions. Taylor series expansion or polynomial fitting is used to obtain approximations to the derivatives of the variables.

The finite element method
According to this method, the domain is divided into a set of discrete volumes or finite elements that are usually tetrahedral or hexahedral. The finite element technique makes use of weighting functions (usually linear), which are multiplied with the equations before their integration over the element area. In the simplest finite element methods, the continuity of the solution across element boundaries is guaranteed in the solution, which is approximated by a linear shape function within each element, and then the approach is substituted into the weighted integral of the conservation law. These functions can be derived from the values at the corners of the elements. The equations to be solved are constructed by the requirement that the integral is zero with respect to each nodal value. So, this corresponds to the solution with the minimum residual. As a result of the above, a set of linear algebraic equations is generated.

The finite volume method
The finite volume method applies the integral form of the conservation equations as its starting point to each control volume in which the domain is divided. Shape functions are used to describe the variation of the parameters over a control volume, which is formed around each node by joining the centroids of the elements.

Detailed description of the finite volume method: The method is based on Patankar's finite volume method (Patankar, 1980). The differential equation of heat conduction is solved using the finite difference technique, but the definition of space has been done according to the finite element method that is more flexible for complex geometries. The three-dimensional heat conduction equation is:

$$\frac{\partial T}{\partial t} = \frac{\lambda}{\rho c} \nabla^2 T \tag{11}$$

where λ is the thermal conductivity (W m^{-1}K^{-1}), ρ the density (kg m^{-3}) and c the specific heat of the soil (J kg^{-1}K^{-1}). Equation (11) depicts the heat diffusion formulation in solids where $\lambda/\rho c$ is denoted as the heat diffusion coefficient.

The one-dimensional steady-state heat conduction equation is:

$$\frac{d}{dx}\left(\lambda\frac{dT}{dx}\right) = 0 \tag{12}$$

For the definition of cells to construct the grid, an illustrative Figure 7.2 is given below.

If dT/dx is defined only at the points W, P and E, or if T is constant over the control volumes surrounding these points, then the distribution of the gradient of temperature along the x-axis is given in Figure 7.3. Therefore, dT/dx is equal to zero within the control volume. If one considers the temperature profile in Figure 7.4 to derive the profile of dT/dx for one dimension, then this is constant within a cell and it can be defined throughout the area between two cells.

Integrating equation (12) over the entire control volume of the cell gives:

$$\left(\lambda\frac{dT}{dx}\right)e - \left(\lambda\frac{dT}{dx}\right)w = 0 \tag{13}$$

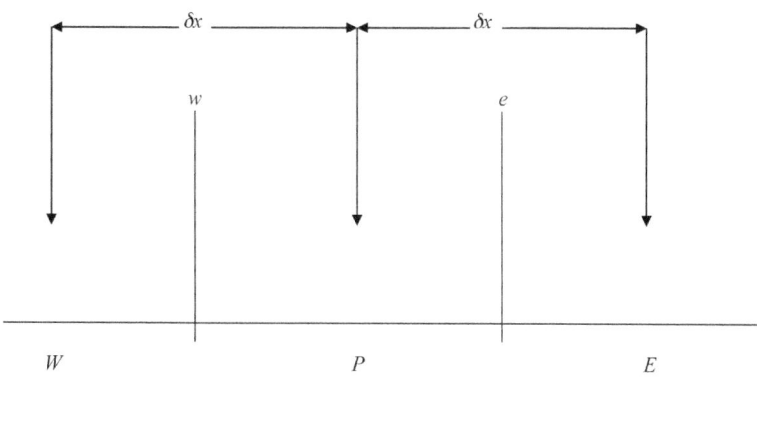

Figure 7.2
Definition of the cell in one dimension, where *w* and *e* are the boundaries

Figure 7.3
Temperature profile where
***dT/dx* is defined only at the**
inside area of the cells *W*, *P*
and *E*, and not at the
control-volume faces

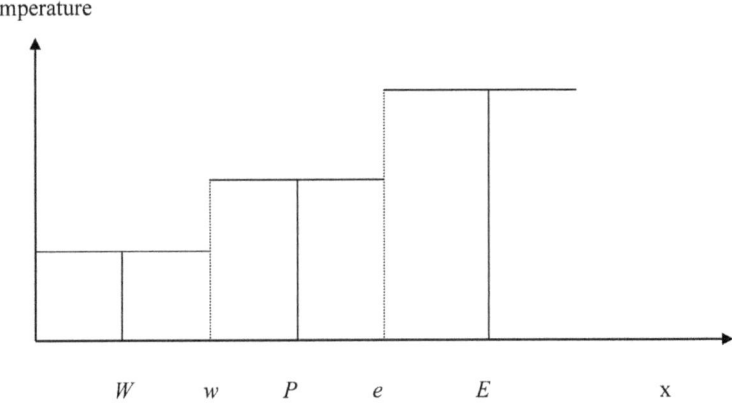

Temperature

Figure 7.4
Temperature profile where
***dT/dx* is defined at every**
point within the cell

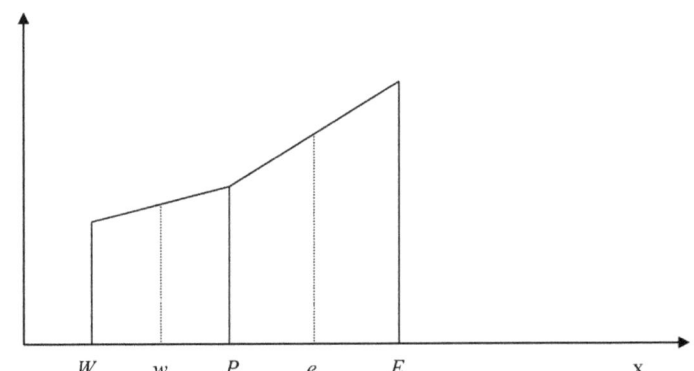

Temperature

After the substitution of dT/dx the steady-state equation (13) is:

$$\frac{\lambda_e\left(T_E - T_P\right)}{(\partial x)_e} - \frac{\lambda_w\left(T_W - T_P\right)}{(\partial x)_w} = 0 \tag{14}$$

If time is taken under consideration we have the equation:

$$\frac{\partial T}{\partial t} = \frac{1}{\rho c}\frac{\partial}{\partial x}\left(\lambda\frac{\partial T}{\partial x}\right) \tag{15}$$

The integration of the above equation over the control volume of the cell leads to:

$$\rho c \int_w^e \int_t^{t+\Delta t} \frac{\partial T}{\partial t} dt\, dx = \int_t^{t+\Delta t} \int_w^e \left(\lambda\frac{\partial T}{\partial t}\right) dx\, dt \tag{16}$$

But

$$\rho c \int_w^e \int_t^{t+\Delta t} \frac{\partial T}{\partial t} \, dt \, dx = \rho c \Delta x \left(T_P^1 - T_P^0 \right) \tag{17}$$

where T_P^1 and T_P^0 are the temperature at point P, at time levels $t + \Delta t$ and t, respectively. Combining the equations (14) and (17) we have:

$$\rho c \Delta x \left(T_P^1 - T_P^0 \right) = \int_t^{t+\Delta t} \left(\frac{\lambda_e (T_E - T_P)}{(\partial x)_e} - \frac{\lambda_w (T_W - T_P)}{(\partial x)_w} \right) dt \tag{18}$$

After the introduction of a weighting factor f between 0 and 1, the time integration results in:

$$\int_t^{t+\Delta t} T_P \, dt = \left(f T_P^1 + (1-f) T_P^0 \right) \Delta t \tag{19}$$

Thus, the algebraic equation for the one-dimensional situation is:

$$a_P T_P^1 = a_E \left(f T_P^1 + (1-f) T_P^0 \right) + a_W \left(f T_W^1 + (1-f) T_W^0 \right)$$
$$+ \left(a_P^0 - (1-f) a_E - (1-f) a_W \right) T_P^0 \tag{20}$$

where:

$$a_E = \frac{\lambda_E}{(\delta x)_E} \tag{21}$$

$$a_W = \frac{\lambda_W}{(\delta x)_W} \tag{22}$$

$$a_P^0 = \frac{\rho c \Delta x}{\Delta t} \tag{23}$$

$$a_P = f a_E + f a_W + a_P^0 \tag{24}$$

So, the three-dimensional case is:

$$a_P T_P^1 = a_E \left(f T_E^1 + (1-f) T_E^0 \right) + a_W \left(f I_W^1 + (1-f) T_W^0 \right)$$
$$+ a_N \left(f T_N^1 + (1-f) T_N^0 \right) + a_S \left(f T_S^1 + (1-f) T_S^0 \right)$$
$$+ a_T \left(f T_T^1 + (1-f) T_T^0 \right) + a_B \left(f T_B^1 + (1-f) T_B^0 \right)$$
$$+ \left(a_P^0 - (1-f) a_E - (1-f) a_W - (1-f) a_N - (1-f) a_S - (1-f) a_T - (1-f) a_B \right) T_P^0 \tag{25}$$

where:

$$a_W = \frac{\lambda_w \Delta y \Delta z}{(\delta x)_w} \tag{26}$$

$$a_E = \frac{\lambda_e \Delta y \Delta z}{(\delta x)_e} \tag{27}$$

$$a_N = \frac{\lambda_n \Delta x \Delta z}{(\delta y)_n} \qquad (28)$$

$$a_S = \frac{\lambda_s \Delta x \Delta z}{(\delta y)_s} \qquad (29)$$

$$a_T = \frac{\lambda_t \Delta y \Delta x}{(\delta z)_t} \qquad (30)$$

$$a_T = \frac{\lambda_t \Delta y \Delta x}{(\delta z)_t} \qquad (31)$$

$$a_P^o = \frac{\rho c \Delta x \Delta y \Delta z}{\Delta t} \qquad (32)$$

$$a_P = fa_E + fa_W + fa_N + fa_S + fa_T + fa_B + a_P^o \qquad (33)$$

Note that, if f equals to 0, then the scheme is explicit (i.e. dependence only on the temperature at the previous step) and if f is equal to 1, then the scheme is fully implicit (i.e. the temperature at the present time step depends on the present nodal temperatures of the nodes that surround the specific node). Specifically, when referring to implicit schemes, it is meant that the system of algebraic equations can be solved either by an *iterative method* (e.g. the Gauss-Seidel method) or by direct methods such as *Gauss elimination*. In the present finite volume method, the above system of equations is solved by iterations. However, the stability of the explicit scheme depends on the size of the time step for uniform conductivity according to the criterion:

$$\Delta t < \frac{\rho c (\Delta x)^2}{2\lambda} \qquad (34)$$

where ρ is the density (kg m^{-3}), c the specific heat capacity (J kg^{-1} K^{-1}) and λ is the thermal conductivity (W m^{-1} K^{-1}). The implicit scheme is unconditionally stable scheme because of its nature.

After the heat conduction equation is solved numerically, it is not possible to calculate the local truncation error because the analytical solution is not known. For these cases it is proposed that a *relative error E* during an iterative process should be given as:

$$E_i = \sqrt{\frac{(T_i^{new} - T_i^{old})}{(T_i^{new})^2}} \qquad (35)$$

where i stands for summation over the nodes that compose the grid and T^{new} and T^{old} are the new and the old values of temperatures calculated. The local relative error in equation (35) is calculated during the iterative process and the time stepping terminates when the criterion $E_i < E_s$ is achieved, where E_s is a predefined tolerance for the iterative procedure.

Air flow domain

Air flow domain is divided into the fluid structure interaction at surfaces (viscous effects) and the free slip flow at distance from earth's surface (boundary layer). Near a rough surface, the viscous effects on the transport processes are large while flow variables vary within the boundary-layer region. In the viscous sublayer, the molecular viscosity plays a dominant role in momentum and heat transfer. Further away from the wall, in the logarithmic layer, turbulence dominates the mixing process.

Assuming that the logarithmic profile reasonably approximates the velocity distribution near the earth's surface, it provides a means to numerically compute the fluid shear stress (SS) as a function of the velocity at a given distance from the surface. Low turbulent Reynolds numbers (Rott, 1990) occur near surfaces that resolve the details of the boundary layer profile. Turbulence models based on the SS are suitable for a low Reynolds number. The density difference is evaluated using the Boussinesq approximation (1897).

The temperature of a fluid in a heated pipe is affected by convection due to the solid-air interface, and due to the air-air interaction. Furthermore, temperature is also diffused inside the air mixture. For a steady-state problem, with the absence of sources, a differential equation governing the temperature will definitely express a balance between convection and diffusion.

The momentum equation for a Newtonian fluid (Potter and Wiggert, 1998) can be written as a generic differential equation with the dependent variable denoted by T:

$$\frac{\partial T}{\partial t} + \nabla \cdot (\vec{u}T) = \nabla \cdot (D\nabla T) + S \qquad (36)$$

| Transient term | Convection term | Diffusion term | Source term |

Where

$\vec{u}$ is the average velocity that the quantity is moving in advection flux.

$D = k/\rho_c$ (k is the thernal conductivity divided by density ρ and specific heat capacity c) is the thermal diffusivity due to heat transport,

The transient term accounts for the accumulation of T in the concerned control volume, the convection term, accounts for the transport of T due to the existence of the velocity field (note the velocity $\vec{u}$ multiplying T), the diffusion term, accounts for the transport of T due to its gradients and the source term, accounts for any sources or sinks that either create or destroy T. Any extra terms that cannot be cast into the convection or diffusion terms are considered as source terms S. For heat transport, $S>0$ might occur if thermal energy is being generated by friction within the viscous sublayer.

In a common situation of an open urban space, the diffusion coefficient might be constant, there are no sources or sinks, and the velocity field describes an incompressible flow (i.e. it has zero divergence). Then the formula simplifies to:

$$\frac{\partial T}{\partial t} - D\nabla^2 T - \vec{u} \cdot \nabla T \qquad (37)$$

The stationary convection–diffusion equation describes the steady-state behavior of a convective-diffusive system. In steady-state, $\partial T/\partial t = 0$, so the formula is:

$$0 = D\nabla^2 T - \vec{u} \cdot \nabla T \tag{38}$$

Unlike the conduction equation, a numerical solution for the convection-diffusion equation has to deal with the convection part of the governing equation in addition to diffusion. Equation (37) is solved by iterative processes in computational fluid dynamics models and techniques. Recently, the convection–diffusion equation has been solved for the bioclimatic rehabilitation of an open urban space (Zoras, 2013).

When heat is added to air and the fluid density varies with temperature, a flow can be induced due to the force of gravity acting on the density variations. Such buoyancy-driven flows are termed 'natural-convection'. In large, open areas where natural-convection flows may be assumed, the Boussinesq approximation could specify the fluid density as a function of temperature. This approximation assumes very small density differences within the air and treats density as a constant value in all solved equations. However, this method leaves gravity to make weight different between the heat exchanger's output air and the ambient air.

It is very common in open spaces to assume turbulent mixing of heat in ambient air. At these cases, particles exhibit additional transverse motion which enhances the rate of energy and momentum exchange between them, thus increasing the heat transfer and the friction coefficient. The Shear Stress Turbulence (SST) model (Menter, 1993; 1994) is a two-equation eddy-viscosity model which has become very popular. The use of a k-ω formulation in the inner parts of the boundary layer makes the model directly usable all the way down to the wall through the viscous sublayer with low Reynolds numbers. The SST formulation also switches to a k-ε behaviour in the free-stream and thereby avoids the common k-ω problem that the model is too sensitive to the inlet free-stream turbulence properties.

The numerical simulation of a ground heat exchanger operating for a passive house ventilation system was recently presented (Flaga-Maryanczyk et al., 2014). The simulation of the earth-contact domain with the buried tube was carried out by a computational fluid dynamics (CFD) software. This way the overall performance of the underground system was studied in relation to soil temperature performance. Similarly, CFD approach has been applied for the evaluation of the above ground thermal conditions in open spaces in two cities (Zoras et al., 2014; Dimoudi et al., 2014)

Earth-to-air heat exchanger (EAHE) – Design considerations

Design characteristics

Ground cooling can be utilised with the application of earth-to-air heat exchangers (EAHXs) at a certain depth inside the ground. The aim is to circulate the hot ambient air through pipes positioned in the ground and cool it at a temperature close to earth temperature and thus, provide air with 1) lower temperature at selected outdoor areas and 2) increased air speed at leeward areas, thus, improving thermal conditions at outdoor areas and mitigating the urban heat island effect.

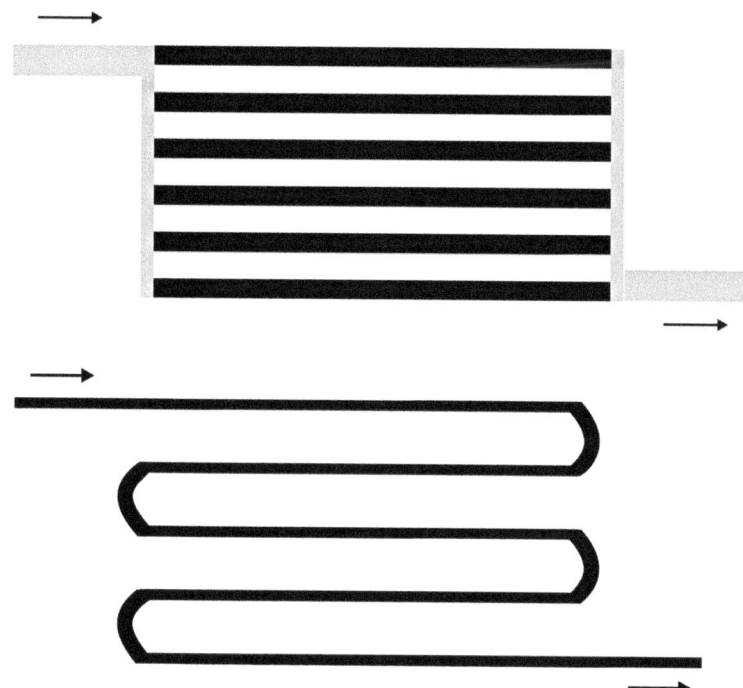

Figure 7.5
Illustration of typical earth
pipes layout

EAHXs typically consist of a network of pipes, with an inlet's haft with filters and an exit pipe, raised at certain height above ground. The network of pipes is placed horizontally or vertically, either in parallel (array), in series, in serpentine (spiral) (Figure 7.5a and 7.5b). Normally, pipe arrays are used to provide higher air flows (Mihalakakou *et al.*, 1994c).

Air circulates through the pipes, usually with the aid of a fan, which is cooled in summer and heated in winter. The exit of the pipes should be at a certain height, e.g. higher than 2.0 m, in order to avoid direct contact of the air with the human face.

The pipes should be seamless and made of steel, aluminium, copper, plastic [polyethylene (PE), Polyvinyl chloride (PVC), polypropylene (PP)], plastic-coated metal pipes (Peretti *et al.*, 2014). The thermal conductivity of the pipe material does not significantly influence the overall heat exchange and, thus, has little influence on summer and winter performance, and the critical parameters are the length and depth of pipes (Badescu, 2007).

Earth pipes can be contaminated with fungi and bacteria during summer operation when warm, humid air condenses at the cool pipe surface. It is thus recommended that pipes should be constructed with a decline of 2.0° in order to prevent water accumulation (Pfafferott *et al.*, 2007). Filters should be used at the inlet of the pipes in order to clean the incoming air and prevent fouling while periodical cleaning – recommended every 2 years using a wet cleaning technique – and filter change ensure a clean interior (Pfafferott *et al.*, 2007). Monitoring of several EAHX systems in buildings have shown that the number of microbes in the outlet air is not higher than the inlet air, even without any cleaning for several

Figure 7.6
Detail of the construction
of the earth pipes, showing
the support of pipes on
horizontal wooden stripes
and the rope guide for the
cleaning system

years of operation (Flückinger *et al.*, 1997 in Pfafferott *et al.*, 2007). Figure of construction details are shown in Figure 7.6.

Air circulation inside the pipes is obtained with a centrifugal fan, fitted at the inlet or exit of pipes. The fan should be of a small size, with low noise in order to avoid noise pollution in the area.

Design parameters

The main parameters that should be considered at the design of an earth-to-air heat exchanger are the:

- mass flow rate
- desired exit air temperature
- pipe diameter
- pipe length
- depth of pipes
- pipes spacing

The boundary conditions are the climatic conditions in the area that define the ambient air temperature, which is the inlet air in the system, and the ground temperature. The ground temperature is related to environmental conditions, soil composition, soil thermal characteristics and its water content.

The design requirements of the system set the magnitude of the mass flow rate and of the exit air temperature. These are achieved with a combination of all the other parameters: pipe diameter, pipe length, depth of pipes and pipes spacing.

High *mass flow rate* and, consequently, high air velocities in the pipe lead to smaller changes of the air temperature inside the pipe (reduction during the cooling period or increase during the heating period) and, thus, are not energy efficient. The size of the fan should be chosen thus, the air velocity speed through the pipes should range between 5.0 to 8.0 m s^{-1} (Argiriou, 1996) while speeds of about 0.7 to 0.8 m s^{-1} were chosen for applications of earth pipes at outdoor spaces (Gaitani *et al.*, 2011; Fintikakis *et al.*, 2011; Santamouris *et al.*, 2012).

The *diameter* of the pipes usually ranges from 0.10 to 0.30 m (Dimoudi, 1996). The smaller diameter pipes have smaller earth contact surface and require more energy to move the air. Larger tubes offer a larger contact of the air with the earth surface, permitting much higher air volumes to be transported through the pipe with a slower airflow.

The optical pipe *length* depends on climate. Increase of pipes' length cause outlet air temperature decrease during the cooling period, which means that the systems cooling potential increases (Michalakakou *et al.*, 1996). The length of the pipes ranges from 10.0 to 60.0 m (Dimoudi, 1996) and it is reported that there are no significant advantages in using pipe length over 70.0 m (Lee and Strand, 2008).

In general, pipe dimensions depend on the desired cooling performance.

Increase of the *depth* of the earth pipes results to higher cooling/heating capacity of the system but it is also related to higher excavation costs. The recommended pipe depth is greater than 1.5 m, as for depths greater than 1.5 m the summer ground temperature is much lower than average ambient temperature (Mihalakakou *et al.*, 1995; Argiriou, 1996). As ground temperatures at a depth of 3.0 m present an almost constant low temperature, close to 18.0°C (Mihalakakou *et al.*, 1992), the depth of earth pipes usually ranges between 1.5 to 3.0 m below ground surface.

The *spacing* between the pipes should be kept approximately 1.0 m from each other in order to minimise thermal interaction (Peretti *et al.*, 2013), as greater spacing was not found to bring extra benefits (Zimmermann and Remund, 2001).

In regards to the *soil composition*, it is very important to ensure the highest thermal contact of the pipe with the surrounding ground. It is thus, recommended that soil is closely packed around the pipe, with at about 0.05 m thick layer; sand or clay are recommended as they have good thermal conductivity and due to their dimensions, they fill the air gaps that would have been formed (Argiriou, 1996; Ascione *et al.*, 2011).

The main design parameters of EAHE are summarised at Table 7.1.

Design examples

Among other heat island mitigation techniques, earth tubes were proposed for bioclimatic redevelopment of outdoor spaces (Gaitani *et al.*, 2011; Fintikakis *et al.*, 2011; Santamouris *et al.*, 2012).

A system of earth-to-air heat exchangers, together with other heat island mitigation techniques (use of cool materials in pavements, green spaces, extensive shading), was proposed in a square, Messolongiou Square, which is situated in a densely built and populated area in the centre of Athens (Gaitani *et al.*, 2011).

Six pipes made of PVC, with 0.20 m diameter, 30.0 m long, positioned at 3.0 m depth were used. The pipes are placed horizontally in a thick layer of sand to improve the heat transfer between the soil and the pipe. Fans in the entrance of each pipe are used to circulate the air through the pipes with a speed close to 7.0 m s^{-1}. The cool air from the pipes is transferred to the ambient air through specially designed outlets placed at 2.5 m height (Figure 7.7).

Table 7.1 Main design parameters of EAHE

	Material	Features	Function
Air inlet			
Air intake tower	Stainless steel	Height: ~2m High efficiency filter	Introduce and filter ambient air
Pipe			
Seamless pipes	Steel Aluminum Plastic: PE, PVC, PP[a] Copper[b]	Resistance coeff, wall roughness, diameter, length, and burying depth. Pipes should be anticorrosive, structurally stable and accessible for inspection and cleaning	Transport air and exchange heat with soil
Stony pipes	Concrete Brick		
Fan			
Air intake fan		Diameter, height and filter type	Take air from outdoors
Exhaust fan			Remove exhaust air
Condensation management		Two types: integrated and standalone	Drain condensation and expel water

[a]   The most widely used today
[b]   Early applications

Source: Adapted from Peretti *et al.*, 2013

The thermal performance of the earth pipes was simulated using the validated model described by Mihalakakou *et al.* (1994b). Simulations carried out for different inlet temperatures shown that during the day period, the exit air temperature is about 6.0–10.0°C lower than the inlet one.

Simulations performed for the whole square showed that the earth pipes contribute to a decrease of the average summer ambient temperature of up to 0.4°C in the square area. Simulations performed for the winter period have shown that for ambient temperatures around 5.0°C, the exit temperature from the earth pipes is close to 12.5°C. It was concluded that, as the use of the exchangers requires some electric energy to run the fans, their use may be limited during the very cloudy days (Gaitani *et al.*, 2011).

A system with earth pipes together with other heat island mitigation techniques (cool pavement materials, high size trees, shading, traffic reduction) were proposed in an urban area in the historic centre of Tirana, Albania (Fintikakis *et al.*, 2011).

Earth pipes made of PVC, 30.0 m long, with a diameter of around 0.25 m were positioned at a depth close to 3,0 m. In total, 15 pipes were proposed to be installed horizontally in the area, positioned in a thick layer of sand with a distance between each other close to 5.0 m. Fans are placed in the entrance of each pipe to circulate the air through the pipes with a speed close to 8.0 m s^{-1}. The air was extracted to the external environment at a height close to 2.1 m through specially decorated vertical tubes (Figure 7.8).

The model proposed by Mihalakakou, *et al.*, (1994b) was used to simulate the dynamic performance of the earth pipes system. It was found that the air temperature at the exit of the pipes, during the summer period, was almost

Figure 7.7 Section of the square showing the full length and the inlet/outlet of earth pipes

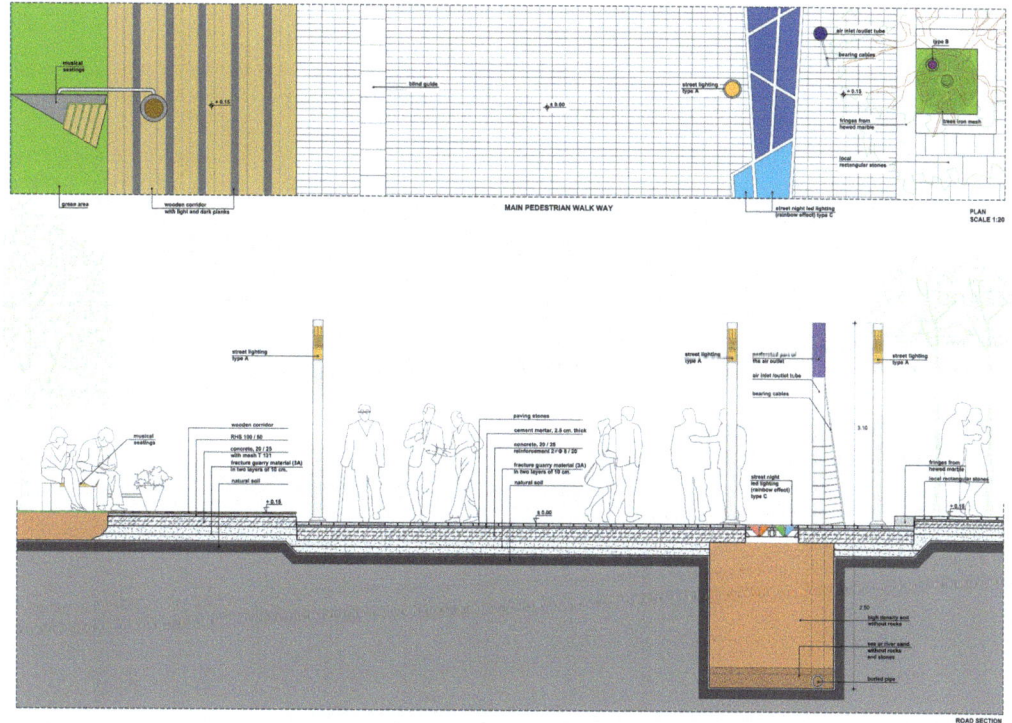

5.0–9.0°C lower than the inlet ones, varying as a function of the inlet temperature and the type of the ground cover, while during the winter period, the temperature increase was between 3.0°C and 5.0°C.

For the whole area, the redevelopment measures result at a maximum temperature drop close to 3.0°C, while the earth pipes contribute up to about 0.7°C. The earth pipes contribute to the decrease of the average ambient temperature in the square and are not found to have a local impact, as the air is quickly diffused from the pipe exits (Fintikakis et al., 2011).

Figure 7.8
Section of the intervention area showing the earth pipes and their inlet/outlets (Architectural design made by Architect N. Fintikakis (Greece) and his office "SYNTHESIS AND RESEARCH LTD")

Conclusions

Earth cooling techniques have been proven to be among the effective urban heat island mitigation techniques. The design of earth cooling techniques, in open spaces, utilises tools and techniques that have been widely analysed and discussed in literature. It is also noted that the implementation of ground heat exchangers has been studied for more than two decades, mainly for buildings. In recent years, a limited number of their applications in outdoor spaces was demonstrated. The design of earth cooling systems to improve ambient urban environment is claimed to be successful if the right combination of tools and methods is used to determine system's performance at urban scale. This chapter presented the most promising tools and techniques to design such systems in terms of acceptance and ease of use. Future research work should be focused on integrated tools that could simulate all the related heat transfer processes in outdoor ground-heat exchangers at urban scale.

References

Adjali, M. H., Davies, M. and Littler, J. (1998a) Earth-contact heat flows: Review and application of design guidance predictions, *Building Services Engineering Research Technology*, vol 19, no 3, pp 111–121.

Adjali, M. H., Davies, M. and Littler, J. (1998b) Three-dimensional earth-contact heat flows: a comparison of simulated and measured data for a buried structure, *Renewable Energy*, vol 15, no 14, pp 356–359.

Argiriou, A. (1996) Ground Cooling in *Passive Cooling of Buildings*, Santamouris M. and Assimakopoulos D. (eds.), Earthscan, London.

Ascione, F., Bellia, I. and Minichiello, F. (2011) Earth-to-air heat exchangers for Italian climates, *Renewable Energy*, vol 36, pp 2177–2188.

Badescu, V. (2007) Simple and accurate model for the ground heat exchanger of a passive house. *Renewable Energy*, vol 32, pp 845–855.

Boussinesq, J. (1897) *Théorie de l'écoulement tourbillonnant et tumultueux des liquides dans les lits rectilignes a grande section*, vol 1, Gauthier-Villars, Paris.

Carslaw, H. S. and Jaeger, J. C. (1959) *Conduction of Heat in Solids*, Clarendon Press, Oxford.

Chartered Institution of Building Services Engineers (CIBSE). (1980) *CIBSE Guide Section A3*, CIBSE, London.

Davies, M., Tindale, A. and Littler, J. (1995) Importance of multi-dimensional conductive heat flows in and around buildings, *Building Services Engineering Research Technology*, vol 16, no 2, pp 83–90.

Dimoudi, A. (1996) Passive Cooling of Buildings in *Passive Cooling of Buildings*, Santamouris M. and Assimakopoulos D. (eds.), Earthscan, London.

Dimoudi, A., Zoras, S., Kantzioura, A., Stogiannou, X., Kosmopoulos, P., Pallas, C. (2014) Use of cool materials and other bioclimatic interventions in outdoor places in order to mitigate the urban heat island in a medium size city in Greece, *Sustainable Cities and Society*, vol 13, pp 89–96.

Fintikakis, N., Gaitani, N., Santamouris, M., Assimakopoulos, M., Assimakopoulos, D.N., Fintikaki, M., *et al.*, (2011) Bioclimatic design of open public spaces in the historic centre of Tirana, Albania, *Sustainable Cities and Society*, vol 1, pp 54–62.

Flaga-Maryanczyk, A., Schnotale, J., Radon, J., Was, K. (2014) Experimental measurements and CFD simulation of a ground source heat exchanger operating at a cold climate for a passive house ventilation system, *Energy and Buildings*, vol 68(A), pp 562–570.

Flückinger, B., Wanner, H. and Lürthy, P. (1997) Mikrobielle Untersuchungen von Luftannsaug-Erdregistern, Institut fur Hygiene and Arbeitsphysiologie, ETH Zurich.

Gaitani N., Spanou, A., Saliari, M., Synnefa, A., Vassilakopoulou, K., Papadopoulou, K., Pavlou, K., Santamouris, M., Papaioannou, M., Lagoudaki, A. (2011) Improving the microclimate in urban areas: a case study in the centre of Athens, *Building Services Engineering Research Technology*, vol 32, no 1, pp 53–71.

Hollmuller, P. and Lachal, B. (2014) Air–soil heat exchangers for heating and cooling of buildings: Design guidelines, potentials and constraints, system integration and global energy balance, *Applied Energy*, vol 119, pp 476–487.

Lee, K.H. and Strand, R.K., (2008) The cooling and heating potential of an earth tube system in buildings, *Energy and Buildings*, vol 40, pp 486–494.

Menter, F. R. (1993) Zonal two equation k-ω turbulence models for aerodynamic flows, AIAA Paper 93–2906.

Menter, F. R. (1994) Two-equation eddy-viscosity turbulence models for engineering applications, *AIAA Journal*, vol 32, no 8, pp 1598–1605.

Mihalakakou, G. (2002) On estimating soil surface temperature profiles, *Energy and Buildings*, vol 34, pp 251–259.

Mihalakakou, G., Lewis, J.O., and Santamouris, M. (1996) The influence of different ground covers on the heating potential of earth-to-air heat exchangers, *Renewable Energy*, vol 7, pp 33–46.

Mihalakakou G., Santamouris, M., and Asimakopoulos, D. (1992) Modeling the earth temperature using multiyear measurements, *Energy and Buildings*, vol 19, pp 1–9.

Mihalakakou, G., Santamouris, M., and Asimakopoulos, D. (1994a) On the use of ground for heat dissipation, *Journal of Energy*, vol 19, no 1, pp 17–25.

Mihalakakou, G., Santamouris, M., and Asimakopoulos, D. (1994b) Modelling the thermal performance of the earth-to-air heat exchangers, *Solar Energy*, vol 53, no 3, pp 301–305.

Mihalakakou G., Santamouris, M., and Asimakopoulos, D. (1994c) On the cooling potential of earth-to-air heat exchangers, *Journal of Energy Conversion and Management*, vol 35, no 5, pp 395–402.

Mihalakakou, G., Santamouris, M., Asimakopoulos, D., Tselepidaki, I. (1995) Parametric prediction of the buried pipes cooling potential for passive cooling applications, *Solar Energy*, vol 55, no 3, pp 163–173.

Mihalakakou, G., Santamouris, M., Lewis, J. O., Asimakopoulos, D. N. (1997) On the application of the energy balance equation to predict ground temperature profiles, *Solar Energy*, vol 60, pp 181–190.

Patankar, S. (1980) Numerical Heat Transfer and Fluid Flow, Hemisphere, New York.

Peretti, C., Zarrella, A., De Carli, M., Zecchin, R. (2013) The design and environmental evaluation of earth-to-air heat exchangers (EAHE). A literature review, *Renewable and Sustainable Energy Reviews*, vol 28, pp 107–116.

Pfafferott J. with Walker-Hertkorn S. and Sanner B. (2007) Ground Cooling: Recent Progress in *Advances in Passive Cooling*, Santamouris M. (ed.), Earthscan, London.

Potter, M. and Wiggert, D.C. (1998) *Fluid Mechanics (Schaum's Series), Schaum's Outlines*, McGraw-Hill, New York.

Rees, S. W., Adjali, M. H., Zhou, Z., Davies, M., Thomas, H. R. (2000) Ground heat transfer effects on the thermal performance of earth-contact structures, *Renewable and Sustainable Energy Reviews*, vol 4, pp 213–265.

Rott, N. (1990) Note on the history of the Reynolds number, *Annual Review of Fluid Mechanics*, vol 22, no 1, pp 1–11.

Santamouris, M., Xirafi, F., Gaitani, N., Spanou, A., Saliari, M., Vassilakopoulou, K. (2012) Improving the microclimate in a dense urban area using experimental and theoretical techniques – The case of Marousi, Athens, *International Journal of Ventilation*, vol 11, no 1, pp 1–16.

Zimmermann M. and Remund S. (2001) Ground coupled air systems in *Low Energy Coupling-Technology Selection and Early Design Guidance*, Bernard N. and Jaunzens D. (eds.), IEA-ECBCS annex 28, subtask 2, report 2, Construction Research Communications.

Zoras, S. (2009) A review of building earth-contact heat transfer, *Advances in Building Energy Research*, vol 3, pp 289–314.

Zoras, S. (2013) Urban environment thermal improvement by the bioclimatic simulation of a populated open urban space in Greece, *International Journal of Ambient Energy*, vol 36, no 4, pp 156–169.

Zoras, S., Tsermentselis, A., Kosmopoulos, P., Dimoudi, A. (2014) Evaluation of the application of cool materials in urban spaces: A case study in the center of Florina, *Sustainable Cities and Society*, vol 13, pp 223–229.

8

URBAN CLIMATE MITIGATION TECHNIQUES

The role of spatial planning

Maria Kaltsa

Kaltsa-Papandreou Architects – Athens, Greece

Introduction

Different cities are shaped by their realities; they reflect cultures and government systems and may face different challenges. Whether success stories or cities in crisis, they all share a common reality when addressing issues of climate change. Climate change is not just about greenhouse gases (GHG); it destroys nature's habitats and biodiversity, induces severe inequalities. It frequently leads the poor and the immigrants to segregation and deprived areas to exclusion, and it intensifies urban decay and causes extreme natural phenomena which erupt more frequently and intensely.

Spatial planners take a bird's eye view of territories. As they hone in and fly closer, their strategies vary and they adapt to a smaller scale. Local characteristics become, thus, more pronounced. They are, in fact, a sort of puppeteer controlling through the frameworks that they define, the ways by which urban systems operate and grow. Though their power to influence policies and strategies on dealing with climate mitigation are profound, they have to consider one significant constraint that poses the greatest threat to their project: namely the human factor. It is the human factor as it pertains to habits, daily behaviour, lifestyle preferences, consumption patterns, connectivity, local identities and cultural traditions that must be taken into account.

In order for spatial planners to effectively address climate change, they must understand cities as systems and realise that growth must meet sustainable considerations. When involved with urban climate mitigation, they are essentially confronted with the task to acknowledge the post-oil world, a world increasingly concerned with the negative aspects of globalisation, and they are expected to bring the global urban model to meet local urban conditions. In ideal terms, sensible planners assess in the best possible way what is to come and thus project an image from the future through planning. They pursue a continuous quest of understanding what are the developing trends imposed by the markets and their impacts, what challenges can be best met in what ways, which negative aspects can be turned to opportunities, how planning sectors must adjust to changes, what forces at work must be coordinated.

The reduction of pollution emissions and environmental degradation can only be achieved through a holistic approach, since effective mitigation strategies demand that, in one way or another, all agents contributing to the negative effects of development, globalisation and environmental erosion must coord-inate and perform better. Conventional energy production, global economic systems coupled with weak local economies, existing patterns of consumption, unsustainable exploitation of natural resources, bad quality of air and soil, the damaged natural cycle of carbon and water, and also serious parameters such as social inequalities, poverty, deprivation of accessibility to services and education, depletion of cultures and identities, all impact climate change.

Being the highest order of planning, spatial planning is highly complex, as it depends on too many factors: geography, topography, geology, size and type of urban regions, their historic evolution, density of habitation, infrastructures, degree of development, political systems, governance styles, labor market and cultural needs, natural habitats, types of ecosystems, availability of natural resources to mention some of them. Spatial planning harmonises all performed actions within the natural and built environment and sets the patterns of human activity. It is, therefore, among the most critical factors influencing urban climate change mitigation and must exercise its tasks efficiently.

Although we live differently in all parts of the world, we are confronted with common problems and challenges relating to climate change; more frequent, extreme occurrences of catastrophic events and other serious symptoms of climatic disorder have made visibly clear that the related social and economic implications are severe.

Climate change will influence future ways of urban living and economic growth globally due to its impact on natural systems and resources. This has made spatial planners in the last decade revise their concepts and strategies in order to combat climate change. The issue has caused abundant preoccupation with the future of urbanism, frequently accompanied with romanticising eco-hysteria and impressive virtual formal or programmatic visual conceptualisation. However, not much has been accomplished that affects critically today's complex realities in specific, realistic ways. Theory and practice do not meet and there is a distance between idealisation and realisation. Visions are very inspiring and necessary and as such they can be radical and exaggerated, but real life scenarios have to comply with much more complex parameters.

The demand for energy will be increasing and the serious side effects of globalisation will continue to persist but still, spatial settlements usually reflect the prevailing socio-economic trends of the past decades. In most nations there is inadequate coordination of sectoral policies in space and lack of policies or regulations. There is non-sustainable use of natural resources, high consumption of conventional energy from polluting sources, mishandling of waste, social segregation and poverty, all of which contribute to environmental damage and severely impact climate change.

On the other hand, there is growing interest in the wealthier developed nations to explore effectively new technological tasks such as info-technologies for sustainable cities that never sleep. The scale and speed of these changes influence patterns of life, work and leisure in space. Observing these social and spatial trends of the new urban world in the information age, sociologist-theoretician Manuel Castells conceptualises in his *Space of flows, space of places* how new experiences and new forms of spatial arrangements have followed the shift from the space of place to the space of flow. He considers intra-metropolitan and inter-metropolitan connectivity a key task of planning, essential in the linking the global to the local, without opposing their operation.

In all cases, sustainable integration of sectoral demands is a key concept today, as spatial planning designs *how* a large scale habitat is structured and performs within the diverse capacities of the entity in which it serves. By structuring a habitat, planning can influence directly the reduction of energy demand and reduction in consumption of natural resources by large consumers, such as cities and peri-urban industries, and dictate 'low carbon' principles while avoiding serious social inequalities. It can dictate the desired degree of a city's compactness or sprawl, regulate use of land on the largest scale, define relations of urban/rural, local/inter-regional/national/trans-border/global, promote synergies and avert conflicts, guide the way to energy efficiency, influence accessibility to networks and resources, balance the types of development which are needed to control exclusion, and suggest mobility patterns, lifestyles and cultures. The habitat's performance then impacts directly its economic, social and environmental sustainability, its competitiveness and growth. Spatial planning must be seen, therefore, as a powerful mechanism for mitigating urban climate change.

What can planning do?

As spatial planning is concerned with the organisation of the physical spatial structure of large regions and provides the framework and directives for large-scale proposals, solutions and regulation, it supersedes in hierarchy urban planning and affects its processes of in essential ways. Regions with different characteristics require spatial development strategies which take into consideration these regional conditions, climates, forms and patterns of built environment and planners involved with urban climate mitigation and low-carbon planning procedures must understand, assess and influence the diverse possibilities about the use of space.

Spatial planning influences environmental management and the dynamics of greenhouse gas generation in cities and can thus play a key role in curbing climate change and its effect on social, economic and environmental imbalances by providing strategies to eliminate conflicts, to harmonise, balance and regulate the impact of all actions in space. Space includes people, their settlements, systems of operation and the natural environment; acts on the natural environment erode its natural processes.

Today, there are many attempts to plan new green settlements to drastically reduce GHG. Most of them engage appropriate green technologies to produce zero-carbon buildings and install solar and other systems to generate heat for their regions. Their advanced infrastructures make use of the logic of a near-perfect metabolism, smart transportation systems and controls to regulate traffic and congestion with advanced mobility technologies. Dynamic research in this field is ongoing. In some rare cases, culture is being regarded as a precious commodity, engaging people to a new lifestyle, liberated from many of the devastating old consumer's habits which relate to the age of carelessness.

Sadly, this age is a past experience for many and a new discovery for many more, the implication of which is felt throughout the planet. The capacity to adapt and cope with the serious impacts of climate change depends upon many factors which determine the performance of any system, smaller or larger. Willingness, wealth of local economies, the relation of global to local, access to technology, availability of the appropriate infrastructures and information technologies, sustainable use of resources, changes in production to consumption habits, promotion of relevant education and information made available to all, development of skills, accountability of the private sector, establishment of institutions and engagement of civil societies and non-governmental bodies, promotion of culture and heritage, and encouragement of diversity and identities, most of which are scarce in poor countries and communities.

Europe has progressed considerably in developing comprehensive strategies to address urban climate mitigation through the discipline of spatial planning. In the context of the European Union, cities are not understood as independent entities but as key actors for all territorial development. In the EU Regional Policy Manual of 2011 *Cities of tomorrow: Challenges, visions, ways forward*, a holistic approach is very well documented, considering the problems and providing directions encouraging the sustainable development of urban environments. Emphasis is given to the relation of global to local for creating resilient and inclusive economies and to what might be more of a European possibility: the trans-regional and transnational collaboration for sharing investments in infrastructures and services or exercising monitoring and controls at a larger territorial scale. Key aspects of the targeted policies are also the projection of existing multinational diversity in all aspects, the avoidance of spatial segregation, rural depopulation and urban drift and the encouragement of people based approaches in governance, which point at adopting a new attitude in spatial planning.

What must planning do?

What types of urban climate mitigation strategies must then be employed by spatial planners and what should their new planning concepts include? Mitigation targets the reduction or avoidance of greenhouse gases and the manipulation of climate and involves different players. Spatial planning harmonises the various expressions of these players, orchestrating their performance according to a preferred story and script. In order for spatial planning to effectively influence environmentally friendly performances, it must engage in a continuous understanding of who the actors are, how they relate and influence each other and their immediate and extended space, and what particular forces are at work in the respective regions within which it operates.

> *Suburbanization led to the separation of city functions, fueling urban sprawl before we became aware of the consequences on climate change and social alienation. Are we are about to repeat the same mistakes, but at a grander and more dramatic scale?*[1]

Controlling the physical expansion of urban settlements as pressures on surrounding land increases and exercising strong controls of land supply and speculative development is a general first measure considered by planners. Integrating new infrastructures and accessibility to transport by different environmentally friendly means is of prime concern, since mobility largely defines the destiny of cities. Sustaining the urban ecosystem in terms of water and energy supply and waste management is essential.

However, combating the impacts of climate change by these means alone cannot produce large-scale results if enough people do not change their habits and patterns of living. Urban environments can provide many essential new amenities, such as smart info networks, but often result in exclusion by way of spatial separation. To produce critically effective results, benefits must be made available to the less privileged and shared by the poorer who cannot afford them. Also, in large cities where entire sections of the population are threatened by exclusion, a mixture of functions and of social groups is necessary. The accommodation of a sustainable growth for the ever-increasing tourist industry and the conservation of cultural heritage – of old and emerging local identities – are necessary to reflect the cities' sense of place.

Today more than ever, simple *responsible* planning must be provided by spatial planners in order to mitigate climate change – planning with a bird's-eye view the destiny of a city and its environment, by employing appropriate assessment, informed judgment, scientific methodologies but also democratic decisions. By incorporating the voices of many players, harmony through planning can contribute effectively to urban climate change mitigation.

The focus on people

Leaving spatial planners and essential aspects of life aside, the focus is still placed today on the production of cleaner or renewable energies, on buildings' technologies, since construction consumes resources, and on integrated mobility technologies; these sectors are accountable for the great chunk of greenhouse emissions. Building technologies relate to urban development as they regard entities comprising building complexes, individual structures, units; they also relate to energy upgrading of buildings, the replacing of their equipment, installation of management systems etc. Spatial planning is not really about *buildings* per se or about developing technologies.

It concerns informed planners today that what is understood and expressed as an eco-crisis is a broader crisis, endemic to how we practice our tasks, live our lives, and it relates to how we produce and consume. It relates to our economic systems, to how we are governed, how we act and perform socially, what culture we cherish; people are now set at the center of an expanding new discussion.

People are so essential to planning and urban climate mitigation that they are regarded as soft infrastructure by U-TT's Alfredo Brillembourg and Hubert Klumpner. Their impact is of equal importance to the material infrastructures the world is mostly preoccupied with. Any approach to urban climate mitigation strategies must encompass people and their things within their specific, integrated environments in a globalised world.

Our epoch is largely defined by the 'R-factor': *Reduce, Reuse, Regenerate,* leading to the adoption of the 'R-planning' logic. The R-factor is associated with high social content. In the recent exhibition of the Deutsches Architekturmuseum (DAM) in Frankfurt, 'Think Global, Build Social' a total focus is on social engagement and the participatory aspect of decision-making for design production. 'R-Urban: Reduce, Reuse, Recycle', a project by Atelier d' architecture autogérée (AAA) for the Municipality of Colombes outside of Paris, includes prototypes for food production (AgroCité), recycling (RecyLab) and cooperatively organised housing (EcoHab), raising awareness among residents about sustainability through workshops.

Globalisation

Globalisation blurs our sense of space, distance, size, time, culture or topography, all of which matter and they need to be seen in revised ways within our understanding of globalisation. Globalised economies have depleted natural environments and can have negative impacts on local economies when local parameters are disregarded. As the current financial and economic crisis reveals, globalisation has made even developed regions vulnerable to external forces. These regions' resistance now largely relies on their local capabilities and the transition to develop more localised sustainable structures. Many are left impotent, and their neglected societies can only perform on a reactive basis. Others, with a higher capacity for action make strategic use of spatial planning in order to define more accountable systems, development patterns and behaviours, while they seek political support and the necessary bodies to implement necessary actions.

Globalisation has raised worldwide concerns over our planet and interest over what is our common home and fate:

> Before opting for a particular setting we must consider our relationship with the universe, our relationship with our species, and find answers to the essential questions. We human beings cannot continue to ignore these questions when we construct our environment, because the way we construct embodies the form in which we believe, and it is first and foremost necessary to be aware of our place on the planet, our place in the universe, in order to transform it accordingly.[2]

Understanding cities in space

The role of cities is very well analyzed in the United Nations Environment Program (UNEP) manual, which provides information on all urban sectors influencing urban climate.[3] Cities must be regarded in the framework of their larger regions, which include sea-space and planning of maritime environments within their influence, since they are drivers of regional development and their smart, sustainable and inclusive growth demands an understanding of a territorial dimension of strategies employed by planners.

Compact city models have gained serious attention and are today favoured since they seem to make best use of technology and resources, they perform and resolve conflicts within the boundaries of used land, adding to their heritage, protecting nature and agricultural land which remains for agricultural production and avoiding wasteful ad hoc developments: 'Integrated recycling networks, methane capture and combined heat and power have relied on ready access to new technologies as well as skilled engineers and installation experts, all of which are easier to access in a compact urban environment'.[4] Transforming the compact city by regenerating its decaying or abandoned parts is an attractive sustainable task, offering dynamic possibilities through the recycling of its used land and buildings.

Outside the compact body of the city, planners use adaptation techniques and place-based approaches to respond to various development conditions, encouraging decentralised models of functions and of governance. Changes in trade, manufacturing and industrial production lead further to the peri-urban transformation of regions. The use of sustainable infrastructures and networks is promoted, along with the reduction of energy consumption from polluting sources and the increase of production of clean energy. In wealthier nations planners are alert, however, to rebound effects of the reduction in order to avoid increasing consumption due to gains. Regarding mobility, expansive motorways are less attractive as they lead to undesired sprawl and environmentally friendly means of public transportation are considered instead:

New roads might be weatherproofed from an engineering standpoint, even taking future climate into account, but they might trigger new human settlements in areas highly exposed to particular impacts of climate change such as coastal zones vulnerable to sea level rise or floodplains (OECD 2009).[5]

As inequalities in urban environment amplify, attention to all human capital is strengthened and valued, as is preserving and reinforcing identities, heritage and cultural assets. Lately, planners encourage governments, the private sector, communities and civil societies to meet in planning, as problems and inequalities deepen and the understanding, assessment and resolution of their conflicts is necessary. New concerns emerge, such as demographics and the needs of the ageing urban society.

However in practice, for all obvious reasons, more interest is expressed over urban development and less over spatial planning. In the global economy, the focus for investment is placed on suitable regions and the speed of change and development of cities has exceeded their formal boundaries, which can no longer reflect their realities. Their impact can be felt in far-reaching diverse ways and it has become the rule that the economic growth does not usually reflect social progress, thus threatening the sustainable future and ability of cities to perform responsibly against climate change. Again, people matter:

The city changes its shapes and forms according to the needs of its inhabitants. In this context, projects are liable to manifest a tendency towards dematerializing; the use of intelligent and programmable materials tends not to separate but to join, to multiply, to let energy flow.[6]

The assemblage of a city is a consolidation of phenomena, exchanges and activities, where the agents are *'human beings and their relation with one another through the flow of matter, energy and information'.*[7]

The optimism about the future of urbanism, especially among imaginative academics, is growing. With the right provisions, technological novelties and infrastructures, cities can become green models for efficient transport, pollution treatment, resource use. Disadvantages are being looked upon as resources and serious problems give way to challenges:

For hundreds of years we designed cities to generate waste. Now it is time that we begin to design waste to regenerate our cities . . . America is the lead creator of waste on the earth, making approximately 30 per cent of the world's trash and tossing out 0.8 US tons (0.72 t) per US citizen per year.'[8]

What types of urban development should spatial planning encourage?

The problems associated with climate change and its impacts fuel diverse theories, fantasies and proposals, which address different tasks and urgencies. The anthology of academic projects keeps growing and presents challenging views of the emerging urbanism or a future one:

Everywhere in nature we find continuous changes that give rise to discontinuous leaps. This is what led R. Thom to formulate catastrophe theory, a mathematical model which envisages a number of patterns of behaviour in systems that experience a sudden change in structure as a result of continuous changes in the parameters which control that structure.[9]

There is much talk about becoming 'smart', developing innovative systems, using smart information and mobility technologies, encouraging smart govern-ance, becoming smart societies:

> *Our city has the job of understanding us, of monitoring the whole complex mesh of interests that leads us to make one decision or another at a particular moment, giving us new places, freeing us from the inconsistency of the habitual.*[10]

Smart cities maybe a target but to accommodate all the necessary prerequisites may not be an easy task, from many points of view. Extensive adaptation of technologies and new systems also leads to diverse models of socio-economic adaptation. Smart integrated systems and infrastructures are considered in nations such as Germany, Singapore or Japan that invest heavily on green tech-nologies, energy, their production, delivery and management but much time will be needed to other nations. And time matters.

For many reasons cities may resist such changes. The quality of their develop-ment depends on many factors, on the types and degrees of pressure imposed upon it and it may necessitate a commitment to educate their societies to respond to new realities or to coordinate all that interact within its region of influence. For most cities, handling their growth and performance sustainably is a major challenge. The responsible task would be to engage smart spatial planners who are more equipped to can target a better quality of life and of services within more sustainable social, economic, environmental and cultural contexts. New mobility systems and networks for public transportation have been very successful in mega-cities throughout the world. Sustainable mobility combined with attractive public transportation contributes considerably to urban climate mitigation and reduces congestion, minimising private transport.

Spatial planners and urban designers contribute most to climate mitigation by creating sustainable environments, by thinking global and acting local, providing place-based, inclusive growth. Large-scale carbon neutral developments cannot be seen as sustainable if they are not inclusive and result in vast, gentrified, walled neighborhoods.

Retrofitting the existing

> *Though they take up only 2 percent of the Earth's surface, cities consume some 75 percent of the world's energy resources and emit around three-quarters of planet's greenhouse gases. Retrofitting existing cities for energy sustainability presents a necessary, but daunting challenge.*[11]

Providing the right type of new infrastructures for urban regions and decreasing the carbon footprint of old ones has a very positive impact in mitigating urban climate change. Buildings are a great contributor to greenhouse gas emissions, their construction consumes resources and generates solid waste. Upgrading old buildings and improving their energy efficiency is a dynamic, emerging new practice globally and the dense, existing urban fabric favours building retrofitting. Quite often, as is the case with vast heritage regions in Europe and elsewhere, recalibration of what exists may be among the few options of climate mitigation planning. In the same spirit, problematic urban environments such as brown fields may be turned into opportunities, as recyclable culture assets with strong identities. Transport planning is essential for their regeneration and of other derelict urban areas.

Optimising use of infrastructures and natural resources

All types of conventional urban infrastructure contribute to GHG emissions and especially in mega-cities and their regions with large informal settlements. Appropriate infrastructures are needed to significantly help in curbing climate change. Cities also do not utilise their natural resources in optimal ways, especially if they happen to cluster within a region belonging to different nations. In order to deal with this problem, the EU has addressed the issue of 'coopetition' (cooperative competition), suggesting the cooperation of such cities as a key to sustainable development. It has also attempted to encourage transnational collaboration through programs such as ESPON and URBACT. Planning the connectivity of urban regions and sharing means that are vital for their sustainable future, such as transportation, energy, water, waste treatment, education and health, industrial zones and trade hubs, research and innovation or sharing policy learning is considered a creative new way for spatial planners.[12]

Urban to rural

In order for cities to release their pressures and cover their needs, they depend increasingly on their surrounding countryside, which is part of their region of influence and of the system within which they operate. Rural areas offer today new opportunities for food and energy production and the pleasures they offer can experienced in sustainable ways. The countryside is also dependent on the city and this inter-dependency can strengthen the region, to the benefit of both parts. Shared services, infrastructures, housing or industrial areas, transport, waste management etc., are incentives to avoid rural depopulation and a forced migration to the city and maintain countryside as an attractive component of the larger urban region.

In balanced territorial developments, the constellations comprising cities, peri-urban spaces which host various sectors and activities and surrounding countryside, are environments that can perform well and contribute to climate change mitigation. However, the rural-urban relationships are complex, and sharing infrastructures and resources within larger networks is not a common practice. Also, the health and balanced development of these constellations depends on their population density; vaster countryside is needed for forests, agriculture and other types of activities.

Infrastructures

Renewable energy: The prospects for the effective mitigating of climate change are not encouraging, though, when the discussion comes to further use of fossil fuel. According to the UN Habitat Global Report of 2011, in the next decade the global urban population is expected to grow by 766 million and 95 per cent of these will be urban residents, mostly in developing nations. Mega-nations which develop and urbanise rapidly open up new coal-powered energy generators every week:

> Of course, the million dollar question will be: 'Do we have enough time? Is this relentless increase in the use of fossil fuels – for at least the next decade – going to affect our planet's climate?' The scientists are very clear on this. They say it will have a devastating effect on the world as we know it. The frightening situation, however, is that it is unlikely that any significant reduction in CO_2 emission is going to happen as long as the developing countries are adding new

fossil fuel generators to their energy infrastructure at the current rate. Just as similar developments were essential for building the western economies over the last 100 years, the same now applies to the developing economies. Their prosperity is dependent on access to cheap energy.[13]

On the other hand, research, development and adaptation of technologies for the production of renewable energies with various methods, has produced impressive results over the years. Energy consumption from green sources is still the most effective way to combat anthropogenic interference with urban climate, as other technologies which can help urban climate mitigation, such as carbon sequestration, are still under development. However, the 'in city' production of renewable energy remains at a very small scale. Spatial planning indicates and regulates all regions which may accommodate the installations for production of alternative energy, within land, water and sea. With renewable energy we curb pollution, encourage diversification and sustainable development, provide energy security and respond to enduring energy poverty.

New mobility: The destiny of a city is largely dependent on and defined by transportation. On the other hand, transport consumes more than half of the planet's fossil fuels and *'contributes around 22 per cent of the world's energy related greenhouse gas emissions'.*[14] Spatial planning can effectively mitigate urban climate change through the proper use of land, adaption of sustainable mobility models and patterns. Actuating smart management systems for existing roads and highways, combined with new systems integrated with energy, non-motorised public and freight transport systems and by encouraging appropriate restricting policies.

Waste: Spatial planning of metropolitan regions has to come to terms with indicating appropriate regions for waste processing, which is probably the most unwanted infrastructure by societies. The volume of garbage and pollution follow developments and the growth of economies. The concentration of pollution and waste that comes even with large compact cities, which are supposed to use their resources and energy more efficiently, remains a major human and environmental health hazard. The hazard is intensified with urban informal settlements. Waste landfills contaminate the soil. Waste contributes about 4 per cent to the world's greenhouse gases and global waste could double in the next 20 years. Billions of tonnes of waste are collected around the world every year: *'Fewer projects or plans have been initiated to address the carbon intensity of the provision of water, sanitation and waste services, or to reduce demand. Outside of the energy sector, there is little evidence that municipalities are linking policies for recycling and reducing waste directly to climate change'.*[15]

Sustainable planning, culture and identity

Culture is a profound bonding agent, necessary for territorial cohesion and contributes to sustainable social and spatial development of communities and regions. It also determines the formal development of structures and the use of public space and in these respects it is used as an essential tool in spatial planning. Understanding cultural diversity is crucial to planning, as is unity in diversity, a crucial factor for obtaining the interest and consent of civil societies. The cultural values of the ecological landscape, whether within cities or in their region of influence and cultural heritage are also factors for economic development, therefore heritage regeneration through its reuse has become a powerful concept of planning.

In the globalised world, cultural landscapes and assets are under serious pressure. The built environment of vast areas in old cities, hosting valuable heritage, has been erased, only to be replaced by what hosts the new models and forms of life. However the conservation and reuse of cultural heritage not only promotes the sustainability of communities in many ways and attracts interest with its aura of social cohesion but also maintains identities which promote the self-esteem of the communities. Today, mainstream civil societies are becoming more and more aware of the negative social implications of monoculture which consumer-oriented urbanism has imposed.

In new or old urban regions which are not associated with cultural heritage, culture and identity remain determining factors for their sustainable develop-ment. The exhilarating pace of transformation in the newly developed econ-omies' built environment, the subsequent changes of lifestyle and behaviour, seeking uniformity, often impose a heavy burden and at times erase old culture to bring in the new. Old areas which are strategically located are gentrified and less privileged populations are pushed to less privileged parts. But urban areas are important fields for social interactions. They frequently host inter-ethnic coexistence of immigrants who support growth and can contribute with interesting new identities, generate profits and sustain better local economies and lives.

In the European Union, development, conservation and management of the cultural heritage is of paramount importance and this largely differentiates Europe from the other continents; Europe essentially is investing in culture to promote competitiveness of its cities internationally.

Agriculture, production distribution and consumption habits

Agriculture and its processes interact with natural resources, biodiversity and ecosystems, whereas the quality of environment and climate change influence agriculture. According to UNEP

> *a temperature increase of 2°C would dramatically reduce the total area available for growing robusta coffee in Uganda, and restrict it to higher altitude areas. Economic modeling studies of farm incomes in India suggest that a 2–3.5°C Increase in temperature would result in a decline in farm net revenues by 9–25 percent.*[16]

Regarding agriculture, planning for climate mitigation requires the sustainable consumption and production of food and that local production-consumption chains are established in urban regions, where agriculture must perform responsibly. Agricultural land must remain a tool for poverty reduction and food security and must be protected from degradation. Environmental degradation leads to inefficient production and requires changes in the patterns of production and the habits of producers, regarding 'how' they produce and market their goods.

The problem is that globalised markets of goods, production patterns, urban-isation, wealth, technology and consumerist habits that disregard local resources and their capacity to renew have led to the adoption of non-sustainable consumption and production patterns. The hard-felt impact on climate change, the environment and its resources, humans, social imbalances, forced migration and human rights is alarming. Seen or unseen wars take place over resources, entire nations depend on others and privileged populations overconsume at the expense of others.

Alongside appropriate planning, consumers must develop environmental and social concerns regarding their consumption habits. To reduce their negative impact on climate change, consumers must be informed about the entire life cycle perspective of goods and their impact, from the use of raw materials, to manufacturing, distribution and disposal. This attitude, encouraged by planning, can redefine local economies, consumption and production patterns.

Governance, civil societies, new models of public-private collaboration

There are many types of governance, so there are many approaches to the issue of climate change. The efficiency of actions regarding climate change adaptation and mitigation of an urban system depend on political attitudes and coalitions, local and national governments, how they formulate their concepts for action, exercise their powers, relate to civil societies and non governmental bodies and what incentives and regulation measures they adopt.

Policy making for urban environments is challenging. In the EU, APRILab has focused in 2010, among other issues, on action-oriented planning and major dilemmas envisioning trade-offs for governance innovation between the extremes of self-organisation and control, addressing: '1. *intervention between control of spatial processes and accommodation of emergent urban change, 2. regulation between instrumentalism and generic normative guidance of self-regulation, 3. investment between supply and demand driven investments'.* It also acknowledged the peculiarities of diversity in urban neighborhoods:

> *Super-Diversity in the neighborhood is a microcosm, mirroring problem constellations at city-scale. But problems of diversity-shaped neighborhoods require locally determined solutions. Thus, municipal policies have increasingly taken a territorial focus in addressing social and integration problems through neighborhood-based initiatives.*[17]

Civil societies are serious actors in promoting awareness, applying pressure and provoking mobilisation of both the government and the private sector, to adopt sustainable performance ethics. Local societies perform well under stress because they know best what their problems are and in what ways they can be addressed in the shadow of crises. Reaching muted audiences and giving voice to many who have no voice, becomes increasingly a sound practice and their contribution is valued: '*Cities are turbulent masses of contradiction, in which exhilaration can turn to frustration at the turn of a corner . . . this book dismisses any notion of control or single viewpoint, and instead allows voices from outside and underneath to bubble up'.*[18]

Concerned governments are reflected in positive, sustainable urban trans-formations and functions enabling in social dialog which leads to a better handling of issues related to climate change mitigation. Smart governments usually reflect their support and use of smart technologies. The private sector, frequently in pursuit of short term profits resulting in environmental degradation, is a powerful player requesting planning permissions. The role of the private sector is becoming increasingly more important as financing and the size of governments shrink in many parts of the world.

Private bodies are asked to provide, finance and manage energy and other vital infrastructure projects, natural resources and services, either in collaboration with the states or under their control. Clearly there are benefits when the private sector and the civil societies are forces at work, however, various complexities and opposing interests make this task very hard to accomplish.

The importance of policy making

When it comes to urban climate mitigation, policy making is one of the most disregarded among the most powerful tools. Spatial planning directs policy makers to several fields which must addressed with policies, specific regulations and controls. The types of policies may reflect larger or smaller scale strategies and spatial entities. National mobility policies targeting large scale impact, provide for the application of desired patterns, regulation and reduction of traffic congestion and reducing emissions which can have an economic impact, such as carbon charges. In order to account for social and other problems with a territorial focus on the neighborhood scale, municipal policies can be locally produced.

Policies institutionalise formal and informal rules, reflecting governance styles and 'ways of doing things'. Their impact is measurable and they develop when intentions meet actions. They can fortify urban systems against external shocks or detect systemic problems and possible threats and can lead to the upgrading of the targeted fields through specific goals. Policies define processes and are indispensable for implementing mitigation strategies by engaging stakeholders, controlling collaborations and regulating global trade, services and virtual entities. Policies must safeguard human interests but also encourage people to adapt changes in thinking and habit.

Land development policies are of particular importance as they define a region's degree of resilience and flexibility. Spatial planning policies must harmonise with a territorial-based logic and any applicable cohesion directives (ex. EU cohesion policies). Directives can be equally powerful as policies, when enforced by funds available from the central governments.

Public policy harnesses shared beliefs and establishes norms that can help territories perform well and make optimal use of their assets. It can help them respond jointly to common challenges, reach critical mass and impact and realise increasing returns by combining activities, exploiting complementarities and synergies between them, thus overcoming divisions stemming from administrative borders. The Commission of the European Communities has communicated the *Green Paper on Territorial Cohesion* in 2008, to address the issues of cohesion of its territories.[19] Many of the described problems which territories face cut across diverse sectors. To tackle problems effectively requires a policy response on a variable geographical scale, involving in some cases cooperation between neighboring local authorities and/or neighboring EU countries.

Territorial cohesion provides sustainable, strong, resilient local systems and economies that can interact more effectively with the global ones. In some member states, metropolitan bodies are created to bring together authorities at different levels to tackle the issues of development that span across their borders, such as public transport, access to healthcare, higher education and training facilities, air quality, waste. Regions such as Eurometropole Lille-Kortrijk-Tournai cross national borders and include cities on both sides of the border. Internal border regions in the EU15 countries have benefited from many years of cohesion policy to improve cross-border cooperation, such as the 'EUREGIO Rhein-Waal', created by German and Dutch local authorities on either side of the frontier to improve among other things the accessibility, quality and efficiency of cross-border health care in the area.

Sustainable growth is inclusive growth

UN Habitat statistics of 2010 report that one third of the world's city population are slum dwellers:

> *Rates of urbanization are higher in developing countries, which are less prepared than developed countries to deal with the resulting impacts. For these regions of the world population growth can act as an acute threat multiplier and significantly exacerbate climate change impacts.*[20]

Climate adaptation is a basic first step to survival of the very poor countries whose deficit accelerates along with climate change. The rising sea levels will destroy entire small island nations and vast coastal areas of others, the lack of food and water or contaminated water, extreme heat waves and the depletion of natural resources will force major populations to migrate to other poor regions, impacting further their problems. UNEP comments that in order to fight poverty and preserve the ecosystems as the foundation of poor people's livelihoods, pro-poor economic growth and environmental sustainability must be placed at the heart of fundamental policies. Responsible spatial planning, accounting disparities which must be confronted, can regulate the use of resources and spatial imbalances, pursuing the development of poor regions which is critical to eradicate their poverty.

In urban regions of more developed nations, spatial planning must give priority to areas near urban centers that suffer economic deprivation and are served less. Increasing inclusion is of paramount importance to avert poverty and the development of informal settlements which generate serious environmental and social problems. Sustainable spatial planning for cities must understand and relate labor market supply to demand and control phenomena of segregation and gentrification and attempt to distribute that the benefits of access to services and to green choices is made available to all:

> *Far more attention is needed from governments and international agencies to building and improving the local institutions that are accountable to and work with urban poor groups. These are also central to the capacity to adapt to climate change. The discussion on the balance needed globally (and within cities) between adaption and mitigation should be informed by the level of the capacity to adapt.*[21]

There are many theories over who pollutes and impacts climate change more: developed or developing nations? Poor nations depend entirely on their access to cheap energy, whatever is available.

> *While urbanization has helped to reduce absolute poverty, the number of people classified as urban poor is on the rise. Urban growth puts pressure on the local environment that disproportionately affects disadvantaged people who live in precarious structures in more vulnerable locations, such as riverbanks and drainage systems, all of which are exposed to flooding, mud slides and other hazards linked to climate change . . . Cities of different wealth levels impact*

the environment differently. As their economies become more prosperous with wider and deeper patterns of consumption, their environmental footprint is increasingly felt at a global level.[22]

Many large developing nations also consume any kind of convenient, cheap energy. Wealthy nations overconsume, produce waste, impose systems of control and dependency through the globalised markets and economies.

We also have to be realistic about the fact that India and China are opening up new coal-powered generators every week. Whatever the rest of the world is doing regarding clean energy is negated by this reality.[23]

The fact is that all must radically reduce green gas emissions and enrich their agenda for responsible planning which should be not be dictated by profit and other interests.

The diversity of urban terrains

Urban terrains have different natural, human and economic potentials and capacities. Depending on their geographic location and continent, their mountainous or other morphological and geographical features, their proximity to coastal zones and natural reserves, their climate, density, culture and economic wealth, they all posses and must maintain their distinct precious characteristics. Some are vulnerable to extreme phenomena resulting from climate change, such as violent storms, floods, surges, water rise, droughts and erosion of soils, extremely high temperatures resulting in extensive fires, depletion of marine and other ecosystems which are valuable to the food chain. Spatial planning must provide risk reduction strategies and resilience to address these hazards.

Reducing consumption and conservation and enhancement of resources and natural systems is a central task for spatial planning. Frequently, water is a central issue when planning for urban regions. It is easily consumed and polluted, it has no real boundaries, its system may be influenced by different regions and nations and its management may require the cooperation of conflicting systems and political bodies. Serious planning with respect to water resources, maritime basins, aquaculture, flood or drought hazards, natural habitats protection and control of pollution may be very complex, due to the fact that rivers, basins and lakes can span along vast terrains. Similar conditions may apply to forestland near urban areas, the degradation and deforestation of which is seriously threatened by the pressures of urbanisation.

Intra-regional and transnational cooperation

Scale is a determinant; sustainable spatial planning for cities should be seen with respect to their dependent regions and a larger territorial collaboration. The connectivity between territories provides multiple benefits in sharing services, infrastructures, communication and mobility networks, production, distribution and business nodes. Intermediate regions between clusters of cities also benefit from networking. Various policies can provide support for such developments, and attractive savings are incentives for collaborations which also benefit climate change.

Territorial cohesion in spatial planning, expressed through cooperation between regions, has strong impact on climate change mitigation without undermining the competitive potential of territories. The EU Territorial Agenda 2020 considers the territorial connectivity as essential for managing and connecting the ecological landscape and the cultural values of regions. However, in many parts of the world where conflict is present this cannot be easily attainable; opportunities are missed to develop shared infrastructures, build collaborative economies, and enforce protection and sustainability measures.

As climate change does not recognise borders, solutions to innumerable urgent problems it creates need to exceed borders. Climate change at a global level requires global cooperation for exercising adaptation and mitigation practices or strategies about resource management such as water: *'Water can also represent a threat. Spatial planning, above all at transnational level, can make an important contribution to the protection of people and the reduction of the risk of flood'.*[24] International initiatives could be excellent tools for urban climate mitigation if they could gain support from key players that are impacting climate change the most: *'During the 2000s, the cities involved in responding to climate change have grown in number and now include cities in the developing world, in part facilitated by the emergence of of new international initiatives'.*[25]

The EU case

Planning at the European scale is seen in the context of sustainable spatial development and reducing GHG emissions is of prime concern. National borders are often an obstacle to the development of European territory as a whole and can restrict its potential for full competitiveness. In the light of the current crisis, intra-regional and transnational cooperation can provide much needed support to regional and national development and economies. Issues of territorial coherence and functional criteria of a geographical nature, such as sharing the same river basin or coastal zone, belonging to the same mountainous area, being crossed by a major transport corridor, demand a broader spatial approach.

Through initiatives of Europeanisation in the European Union, a poly-centric territorial development is favoured with interdependent urban-rural partnerships and integrated spatial planning approaches. The hopes are that in the near future infrastructures, hubs or large projects can be shared transnationally. In pursuit of balanced territorial development, transnational programs were initiated through various efforts such as the Spatial Development Perspective (ESPD) and the ESPON observatory. An urban system with strong partnerships between urban and rural areas redefines the relationship between city and countryside. *'A complex web of large, medium-sized and smaller cities has arisen, which in large parts of Europe form the basis for urbanized spatial structures even in agricultural areas'.*[26]

The trends for change in the EU urban system point at urban networks, making spatial planning an important tool to promote issues of European significance. The *Green Paper on Territorial Cohesion* provides a thorough outline of all critical issues that are to be addressed in EU directives.

> *Territorial cohesion builds bridges between economic effectiveness, social cohesion and ecological balance, putting sustainable development at the heart of policy design . . . The settlement pattern of the EU is unique. There are about 5 000 towns and almost 1 000 cities that spread across Europe, acting as focal points for economic, social and cultural activity. This relatively dense urban network contains few very large cities. In the EU, only 7 per cent of people live*

in cities of over 5 million as against 25 per cent in the US, and only 5 EU cities appear among the 100 largest in the world. This settlement pattern contributes to the quality of life in the EU, both for city dwellers living close to rural areas and those rural residents within easy reach of services.[27]

The *territory* has thus become the new dimension of EU spatial development strategies, leading to the development of policies for larger areas, with strong spatial impact. Territories contain cities and urbanised regions which are very common in parts of central and north-central Europe. Their diversity is to be preserved and enhanced as it is an asset for their sustainable development.

Major problems to be resolved are the continuing urban sprawl, increasing social inequalities and segregation in cities, demographic changes, changes in the function of rural areas which have consequences for the economies and land uses, imbalanced transport, lack of networking, increasing pressure on cultural heritage. Addressing remote or less privileged regions which are deprived from access to adequate energy, education, services, goods and innovation must also be solved. The EU Cohesion Policy, period 2007–2013, supported with funding three principal objectives: convergence of the economically weakest states with stronger economies; improving competitiveness and employment in weaker regions; and strengthening territorial cooperation.

The opportunities and challenges of the polycentric development model are many. It is the basis for better accessibility and organisation of trans-European networks and for inter-regional, cross-border and transnational cooperation between the member states. The urban-rural partnerships have better access to efficient and sustainable infrastructures, services and knowledge. They can share water resource management, common agricultural and environmental policy and promote indigenous development and diverse rural areas.

The EC Cohesion Policy Strategic Guidelines provide good information on the diverse challenges relating to territorial cohesion.[28] INTERREG is the program that challenged administrative boundaries of adjacent EU regions for cross-border cooperation and partnerships regarding transnational energy networks, activities, information exchange and the sharing of experience. It engaged thousands of actors in cooperation projects and tackled spatial planning for large constellations of geographical areas to address issues such as flooding or drought problems.

The INTERREG I Initiative was launched in 1990 with a budget of one billion Euros in order to overcome the disadvantages presented by administrative boundaries of adjacent regions in the emerging Single Market. The INTERREG II Initiative, from 1994 to 1999, had a total budget of EUR 3.5 billion. It continued the cross-border cooperation activities, but also provided funding for transnational energy networks.

From 1997, in the context of the preparation of the ESDP, an additional funding strand ('C') on transnational cooperation to tackle flooding and drought problems and to develop spatial planning for large groupings of geographical areas was introduced with a budget of 440 million Euros. The Community Initiative INTERREG III, in the Structural Funds period 2000 – 2006, had a budget of more than five billion Euros. Three objectives were supported: cross-border cooperation (Strand A); transnational cooperation on spatial planning across large contiguous areas (Strand B); and interregional cooperation to improve the effectiveness of regional development through information exchange and sharing of experience (Strand C). Strand B was most explicitly concerned with spatial planning and the application of the ESDP policy concepts such as polycentric territorial development and urban-rural partnerships. [29]

Europe versus Asia and America

Different growth patterns exist in different continents. Asian cities are experiencing rapid growth, whilst in Europe the cities are no longer growing and *'Links between cities in Europe are considerably more extensive than those on other continents'.*[30] Planners of mega-cities in Asia and America are frequently confronted with different problems than those of planners in Europe.

Mega-cities are key players in the fight against climate change. Even though climate mitigation strategies are more cost effective and can prove more efficient in the more compact ones, according to the 2006 OECD territorial reviews on competitive cities in the global economy, the relationship between income and population size in many such cities becomes negative at around 6–7 million.

A spatial development perspective restricted to individual metropolitan regions is not in line with EU concepts. Europe targets the harmonisation of the global to the local, and favours instead a sustainable development of local entities into a polycentric structure across its territory. However climate change is a global problem and the relating challenges remain common, as are the presented strategies and responses for urban climate change mitigation.

What is to come?

Consumption of fossil fuels, pollution from industries and waste, changes of lifestyle and land use, the destruction of forests and natural habitats and poverty damage the natural cycles such as carbon and water in the planet's balance. Spatial planning can address the reduction of the ecological footprint and provide directions for low-carbon habitation and production processes, and is therefore a precious tool in mitigating climate change. Planning may also curb conflicts of use and impact a variety of other problems, existing, developing or of the future. It can contribute to establishing strong local economies and societies which can function autonomously or by adapting to their environment and culture. By considering the broader territorial context possible, it can guide local and other governments to share sustainable visions for their interdependent regions. But what is to come? Much is predictable and some is not.

The predictable outcome would be a growing awareness and demand for urban sustainability and resilience, comprehensive planning and consideration of all parameters which impact a region's development. Planners will address imbalances by suggesting strategies. They will have to identify the availability of energy demand within territories and allocate where to install new, green energy systems, define transport, trading hubs and corridors, regional systems for agricultural, industrial or manufactured goods, urban development patterns. They will enforce synergies of sectors and technologies, policies, infrastructures, people:

> *The new history will be constructed on the basis of distributed, decentralised systems, by way of operational nodes – people, things, places, territories – that cooperate freely in order to be more efficient . . . As in all mutations, the saturation of the city's vital systems leads to their re-programming on the basis*

of principles that are closer to those of information systems than the simple accumulation of inorganic matter.[31]

The less predictable outcome is its vulnerability to the unpredictable economic forces and shifts in practices that will take place. The stage for spatial planning in the last decades was set by the economy of the globalised world; now there is a need for reconsidering *who* must be the set designer.

What must be done

Cities are complex entities; adjustments in human habits, behaviour and various systems at work are essential in realising meaningful results for mitigating urban climate change. Efforts must be made to reduce their vulnerability to external changes and enhance the capacity of society to adapt to climate change. In view of an emerging world, it is also interesting to address future challenges while responding seriously to the current ones.

Environmental planning, an integral component of spatial planning for urban systems, must harmonise sectoral actions to take place in physical space. It must link networks of macro systems through wide inter-regional, transnational and international cooperation, considering the broader territorial context possible. It must engage governance to share sustainable visions and policies for interdependent regions. The current model of production-services-consumption-markets must be redefined to meet a more sustainable logic. Resilient local economies must be developed and societies which can function autonomously, adapting to their environment and culture.

Urban areas must transform into centers of innovation, use smart transport and car free mobility, accessible nodes with different types of action, sustainable infrastructures and management systems for optimising their sustainable operation. Imbalances, social exclusion, poverty and loss of identities must be seriously addressed. Low densities of new settlements relate to urban sprawl, short term investment returns and profits. Sprawl overconsumes energy, resources, causes extensive development of infrastructures and must be avoided: '*Urban environments with higher density residential and commercial buildings, a well distributed mix of uses and public transport reduce the energy footprint'.*[32]

In addition to technologies, systems, economic and other considerations, spatial planning considers *people*; through its provisions, it can and must promote the much-needed behavioural change which is required in order to critically mitigate climate change.

What must be stressed

It is a central intention of this paper, to stress that what is often overlooked in urban climate mitigation strategies, is the human factor. Climate mitigation goes way beyond the reduction of CO_2 emissions; cities are not just hubs for material and technological innovation but also hubs for life, where daily choices impact humans and their environment.

Radical climate mitigation techniques alone cannot save the planet. A wide-spread, radical change of people's habits must also be addressed, along with a modification of behaviour and a will to realise changes regarding how we produce and consume, how we live. Nature remains larger than life. The respect nature demands and the balance our reliance on nature necessitates can be secured by building sustainably and acting responsibly.

Exercising democratic governance enables spatial planning to safeguard climate by protecting the interests of healthy populations, balanced developments and human rights: *'Cities are uniquely placed to harness their human and environmental potential, guiding urban growth towards greater social and environmental equality'.*[33] A Counsel of Europe Symposium of 2008 was dedicated to the spatial dimension of human rights, focusing on issues of ethics and health relating to the management of territories.[34] But governments and the markets often exercise pressure on planners to divert from sustainable concepts and strategies. Michael Sorkin sees 'distribution' as a defining element of what architecture engages, be it mass, space, materials, privilege, access, meaning, shelter, rights. What he thinks of as wrong regarding architecture in relation with capitalism is *'the unequal aggrandizement of space, convenience, and privilege embodied and represented'.*[35] Substituting the word 'architecture' with 'planning' presents an equally solid case of the argument and indicates these contents of the planning discipline to which much more attention must be paid.

When in crisis, we witness a deregulation of the economic system which impacts the political one, leading – a previously muted – society to the quest of new expressions, dynamics and social demands. As many current crises reveal, current economic practices cannot provide enough jobs for all people in cities, balanced territorial development throughout regions and the world; serious adjustments have to be implemented before pressures can be taken off the natural environment to comfort the planet.

Cities are ever-changing kaleidoscopes, re-figuring their content. Spatial planning is not a sensational practice such as architecture or design; it provides no images and visual references and as such, it is largely liberated from materiality. It can be visionary, connect people through the sharing of visions, and provide for equitable socio-economic development, social cohesion and integration. No matter how many theories and strategies the future of planning will develop,

> *none of the envisioned systems could ever be conceived without a primary condition: an optimistic, collaborative society with strong links to nature . . . reflect on the bold decisions that we as a society will soon be asked to make, in order to change the social and economic structures that determine the way we built our cities.*[36]

Epilogue

Space is the container of all human actions on the planet, spatial planning is therefore influential on both humans and the planet. Sensible planning orchestrates patterns for living better lives through synergies, technologies, policies, infrastructures, and people. As a powerful tool for climate change mitigation, integrated strategies can suggest new patterns of production, consumption and challenge society to adapt to the lifestyle changes which are necessary for an optimistic future. A critical overview of the issues involved in mitigating urban climate change, suggests a necessary shift from low to high coordination of sectors and activities within territories, the capacity of which must be evaluated. Planning decisions must reflect good ethics, concerns for one planet and the oneness of humanity.

Many good practices and positive changes take place today in urban regions, suggested by spatial planners, but their impact is not yet critical; many more must be considered and applied in the near future since *the old form is no longer effective and the new form is not clearly visible yet'.[37]* The relationship of supply and demand will remain a key factor for planning sustainably:

> *In the emerging world of instant and universal connectivity, the intellectual and operational challenges for architects, space planners and facilities managers, who together have the responsibility of articulating the demand side of the supply/demand equation, are fourfold: (1) to be parsimonious in justifying the value of place in an increasingly virtual universe; (2) to invent temporal and behavioral conventions that will respond to enhanced demands for instant accessibility and constant communication; (3) to value the modification and improvement of the existing built fabric as at least equally important to creating the new and (4) to invent the physical fabric that will be needed by emerging Twenty First Century cities and organizations. In this new world my conjecture is that less really will become more.[38]*

This paper reflects some of the less conspicuous but crucially impacting considerations of the spatial planning discipline, leaving aside the political side of systems and economies at work. However, there is urgent need for the rising global super-nations to be amongst the key actors influencing climate change and adopting mitigation strategies. It will be interesting as to see how they will address and identify relevant areas of concern, as this will be a direct reflection of their economic and political ethos. In that respect it is interesting to refer to Peggy Deamer's comment that

> *Vis-à-vis the world economy, India and its BRIC(S) colleagues (Brazil, Russia, India, China – and, since 2010, South Africa) operate on the thesis that China will soon become the dominant global superpower of dominant goods, India of services, and Brazil and Russia of raw materials. While neither a political alliance nor a formal trade association, BRICS is considered to be an attack on the Western-dominated IMF and World Bank. It is anticipated that the BRICS economies will overtake the G7 by 2027. By then, Western tropes of modern-ization will no longer be tropes; they will be the new real thing.[39]*

We have little certainty and can only speculate as to what challenges cities will face in the future. We cannot assess what will be the outcome to populations from upheavals and wars, poverty, extreme shifts in labor markets, production systems, technological achievements and governance styles. In response, we must be alert and anticipate changes as well as possible, structure resilient urban environments and view problems as opportunities for change. Focus must be kept on the powerful human factor, since *'Rediscovering the fragile thread that links physical order to human behaviour will be the main task of this Urban Age, a world where 75 per cent of us will be living in cities'.[40]*

> *Damage must be prevented from happening in the first place, and investing on prevention will produce the highest gains. I see the role of spatial planning for urban environments as that of a naturally healing and preventive medicine: assist a system to recover and forge resilient immunity. Then, it comes down to influencing the urban system and changing some of its bad habits: 'Why fish bodies out of the river when you can stop them jumping off the bridge?'.[41]*

Notes

1   Burdett R., Rode P. 2011. Living in the urban age, in *Living in the endless city*, edited by Ricky Burdett, Deyan Sudjic. London: Phaidon Press Ltd: p.8.

2   Cappelli L. 2010. A world of settings, in *Self sufficient city – Envisioning the habitat of the future*, edited by Lucas Cappelli, Vicente Guallart. Barcelona: IaaC-Institute for Advanced Architecture of Catalonia, Actar: p.22.

3   U.N. 2014. *U.N.E.P. – United Nations Environment Programme.* [Online]. Available:. http://www.unep.org/climatechange/mitigation/Default.aspx [3 January 2014].

4   Stern N., Zengelis D., Rode P. 2011. City solutions to global problems, in *Living in the endless city*, edited by Ricky Burdett, Deyan Sudjic. London: Phaidon Press Ltd: p.345.

5   U.N. 2011. *U.N.E.P. – United Nations Environment Programme.* [Online]. Available: http://www.unep.org/pdf/mainstreaming-cc-adaptation-web.pdf (p. 7). [3 January 2014].

6   IaaC – Institute for Advanced Architecture of Catalonia. 2010. Living landscapes, in *Self sufficient city – Envisioning the habitat of the future*, edited by Lucas Cappelli, Vicente Guallart. Barcelona: IaaC – Institute for Advanced Architecture of Catalonia, Actar: p.42

7   IaaC – Institute for Advanced Architecture of Catalonia. 2010. Social & collaborative, in *Self sufficient city – Envisioning the habitat of the future*, edited by Lucas Cappelli, Vicente Guallart. Barcelona: IaaC – Institute for Advanced Architecture of Catalonia, Actar: p.314.

8   Joachim M., Fessel M. 2014. Rethinking urban landscapes: Self-supported infrastructure, technology and territory, in *Cities for smart environmental and energy futures– Impacts on architecture and technology*, edited by Stamatina Th. Rassia, Panos M. Pardalos. Berlin Heidelberg: Springer: p.26.

9   Mueller W. 2010. The city and me, in *Self sufficient city – Envisioning the habitat of the future*, edited by Lucas Cappelli, Vicente Guallart. Barcelona: IaaC – Institute for Advanced Architecture of Catalonia, Actar: p.12.

10 Cappelli L. 2010. A world of settings, in *Self sufficient city – Envisioning the habitat of the future*, edited by Lucas Cappelli, Vicente Guallart. Barcelona: IaaC – Institute for Advanced Architecture of Catalonia, Actar: p.23.

11 U.N. 2011. U.N.E.P. – United Nations Environment Programme – Integrated Approach for Low Emissions Project Development in the New Town of Boughzoul, Algeria. [Online]. Available: http://www.unep.org/energy/Activities/LowEmissionsProject Development/tabid/79484/Default.aspx. [3 January 2014].

12 E.U. 2011. Cooperation between cities is key to sustainable European urban development, Conclusions – a strengthening of the European urban developing model, in *Cities of Tomorrow*. Luxembourg: Publications Office of the European Union: p.85–89.

13 Budde P. 2014. Smart cities of tomorrow, in *Cities for smart environmental and energy futures – Impacts on architecture and technology*, edited by Stamatina Th. Rassia, Panos M. Pardalos. Berlin Heidelberg: Springer: p.12.

14 Stern N., Zengelis D., Rode P. 2011. City solutions to global problems, in *Living in the endless city*, edited by Ricky Burdett, Deyan Sudjic. London: Phaidon Press Ltd: p.345.

15 U.N.-Habitat – United Nations Human Settlements Programme. 2011. Climate change mitigation responses in urban areas, in *Global report on human settlements 2011, Cities and climate change: Policy directions*. London: Earthscan Ltd: p.29.

16 U.N. 2011. U.N.E.P. – United Nations Environment Programme – Mainstreaming Climate Change Adaptation into Development Planning: A Guide for Practitioners. [Online]. Available: http://www.unep.org/pdf/mainstreaming-cc-adaptation-web.pdf. (p. 8). [3 January 2014].

17 Joint Programming Initiative Urban Europe, *Urban Europe – APRILab.* [Online]. Available: http://jpi-urbaneurope.eu/project-aprilab/. [3 January 2014].

18 Till J. 2012. Responses to Critical Cities Volume 1 and Volume 2, in *Critical Cities Volume 3*, edited by Deepa Naik and Trenton Oldfield. London: Myrdle Court Press.

19 Commission of the European Communities. 2008. *EUR-Lex, Access to European Union law – Green Paper on Territorial Cohesion Turning territorial diversity into strength.* [Online]. Available: http://eur-lex.europa.eu/LexUriServ/LexUriServ.do?uri=COM:2008:0616:FIN:EN:PDF. [3 January 2014].

20 U.N.-Habitat – United Nations Human Settlements Programme. 2011. The impacts of climate change on urban areas, in *Global report on human settlements 2011, Cities and climate change: Policy directions*, edited by UN. London: Earthscan Ltd: p.24.

21 Satterthwaite D. 2011. Surviving in an urban age, in *Living in the endless city*, edited by Ricky Burdett, Deyan Sudjic. London: Phaidon Press Ltd: p.379.

22 Burdett R., Rode P. 2011. Living in the urban age, in *Living in the endless city*, edited by Ricky Burdett, Deyan Sudjic. London: Phaidon Press Ltd: p.11.

23 Budde, P. 2014. Smart cities of tomorrow, in *Cities for smart environmental and energy futures – Impacts on architecture and technology*, edited by Stamatina Th. Rassia, Panos M. Pardalos. Berlin Heidelberg: Springer: p.12.

24 European Commission. 1999. *European Commission – ESDP (European Spatial Development Perspective).* [Online]. European Commission. Available: http://ec.europa.eu/regional_policy/sources/docoffic/official/reports/pdf/sum_en.pdf. (Chapter 3.4.3 Water Resource Management-A special challenge for spatial development)

25 U.N.-Habitat – United Nations Human Settlements Programme. 2011. The impacts of climate change on urban areas, in *Global report on human settlements 2011, Cities and climate change: policy directions.* London: Earthscan Ltd: p.25

26 European Commission. 1999. *European Commission – ESDP (European Spatial Development Perspective).* [Online]. European Commission. Available: http://ec.europa.eu/regional_policy/sources/docoffic/official/reports/pdf/sum_en.pdf. (p.64). [3 January 2014].

27 Commission of the European Communities. 2008. *EUR-Lex, Access to European Union law – Green Paper on Territorial Cohesion, Turning territorial diversity into strength.* [Online]. European Commission. Available: http://eur-lex.europa.eu/LexUriServ/LexUriServ.do?uri=COM:2008:0616:FIN·FN·PDF. [3 January 2014].

28 European Commission. 2005. *European Commission – Cohesion Policy in Support of Growth and Jobs: Community Strategic Guidelines, 2007–2013.* [Online]. European Commission. Available: http://ec.europa.eu/regional_policy/sources/docoffic/2007/osc/050706osc_en.pdf . [3 January 2014].

29 Dühr S., Stead D., Zonneveld W. 2007. Taylor & Francis Online – Planning practice & research, volume 22, issue 3, 2007, Special issue: The Europeanization of spatial planning through territorial cooperation. [Online]. Routledge: p.294. Available: http://www.tandfonline.com/doi/full/10.1080/02697450701688245#.Um9UkeXodrY. (p.294). [3 January 2014].

30 E.U. 2011. Competition as a zero-sum game, in *Cities of tomorrow.* Luxembourg: Publications Office of the European Union: p.22.

31 Guallart V. 2010. Can the planet withstand another 20th century, in *Self sufficient city – Envisioning the habitat of the future*, edited by Lucas Cappelli, Vicente Guallart. Barcelona: IaaC-Institute for Advanced Architecture of Catalonia, Actar: p.8.

32 Burdett R., Rode P. 2011. Living in the urban age, in *Living in the endless city*, edited by Ricky Burdett, Deyan Sudjic. London: Phaidon Press Ltd: p.14.

33 Burdett R., Rode P. 2011. Living in the urban age, in *Living in the endless city*, edited by Ricky Burdett, Deyan Sudjic. London: Phaidon Press Ltd: p.24.

34 2009. European spatial planning and landscape, No 91. International CEMAT Symposium on 'The spatial dimension of human rights: for a new culture of the territory'. Council of Europe.

35 Sorkin M. 2014. Afterword: architecture without capitalism, in *Architecture and capitalism*, edited by Peggy Deamer. London, New York: Routledge: p.218.

36 IaaC – Institute for Advanced Architecture of Catalonia. 2010. Theories & strategies, in *Self sufficient city – Envisioning the habitat of the future*, edited by Lucas Cappelli, Vicente Guallart. Barcelona: IaaC – Institute for Advanced Architecture of Catalonia, Actar: p.371.

37 Nowak W. 2011. Foreword, in *Living in the endless city*, edited by Ricky Burdett, Deyan Sudjic. London: Phaidon Press Ltd: p.6.

38 Duffy F. 2014. Redesigning the relationship between supply and demand – A time and a place for everything, in *Cities for Smart Environmental and Energy Futures – Impacts on Architecture and Technology*, edited by Stamatina Th. Rassia – Panos M. Pardalos. Berlin Heidelberg: Springer: p.5

39 Deamer P. 2014. Context: 1990–2010, in *Architecture and capitalism*, edited by Peggy Deamer. London, New York: Routledge: p.171.

40 Burdett R., Rode P.. 2011. Living in the urban age, in *Living in the endless city*, edited by Ricky Burdett, Deyan Sudjic. London: Phaidon Press Ltd: p.25.

41 Parker B. 2013. Nothing new on the idea of recycling (from a text collage by Reiner de Graaf), in *ARCH+ Journal for Architecture and Urbanism, Deutsches Architekturmuseum No. 211/212*, edited by Arch+. Aachen, Berlin: ARCH+: p.111.

9

URBAN CLIMATE MODELS

Christine Georgatou and Denia Kolokotsa

Technical University of Crete – Chania, Greece

Introduction

During the past few decades, environmental, physical and architectural research has caused a fast evolution in outdoor urban environmental models. Temperature changes resulting by the built-up setting affect people's health and comfort in addition to energy consumption and air quality. For that reason it is significant for the urban planners to learn about air temperature variations between different land-use categories for both extreme situations and during average conditions. [1] Furthermore, it is also of great importance to be able to provide accurate predictions in the longer term for the future situation for the study of climatic conditions when designing a city. The flow of air and the thermal structure of the atmospheric boundary layer (ABL) is defined by the earth's surface. In particular, the surface energy balance, the separation of energy at the surface into diverse kinds, and the roughness of surface establish the temperature and the vertical profiles of wind and temperature in the boundary layer. Urban areas alter the material and aerodynamic character of the surface, greatly affecting the surface energy balance, as well as the dynamic and thermodynamic nature of the boundary layer. These modifications to the local climate are the core topics of urban meteorology and urban climatology. [2] Concerns on negative effects of urbanisation on the environment make the characteristics of urban areas increasingly important in urban planning and building construction, particularly at high density cities. [3] The urban boundary layer (UBL) is the part of the atmosphere in which most of the planet's population now lives, and is one of the most complex and least understood microclimates. Given potential climate change impacts and the requirement to develop cities sustainably, the need for sound modelling and observational tools becomes pressing. [4] Nowadays it is evident that the increase of urban temperatures has a serious impact on the energy demand of buildings by increasing significantly the energy consumption for cooling, while decreasing to some extent the energy consumption for heating. Moreover the urban landscape creates a climate which affects human comfort, air quality and energy consumption. Though, regardless of these facts, climate issues often have low impact on the urban planning process in practice. [5] The air flow in the ABL is characterised by turbulence, which is generated by wind shear (wind is approximately geostrophic at the top of the ABL but zero at the surface). Temperature gradients can either generate or suppress turbulence. [6] It is the layer where nature springs, humans live and all our activities occur and the pollutants disperse affecting the rest of the atmosphere. Furthermore, all meteorological conditions prevail considered for the study or forecasting of the climate changes, and the temperature variations.

During a clear day, the boundary layer can be divided into several sublayers, as shown in Figure 9.1:

1. The roughness sublayer: this is the layer of air in which air flows around individual roughness elements (such as grass, plants, trees or buildings).
2. The surface layer: (formerly known as the constant flux layer) in this layer, typically 100 m thick (or 10 per cent of the depth of the ABL), the winds, temperature and humidity vary rapidly with altitude, and the characteristics of turbulence are affected by the surface. Vertical fluxes of heat and momentum are approximately constant.
3. The well-mixed layer: rising buoyant plumes from the surface layer, and associated turbulence, cause potential temperature and other quantities to be relatively constant with altitude. The earth's rotation becomes important in this layer, and the wind direction veers with height.

The spatial scales main of the UHI study and modelling are:

1. **Mesoscale:** The UHI impact study ranges in a zone 10^4–10^5 m above the city surface. The urban environment is considered as a whole, differentiated from its surrounding rural or forest areas.
2. **Localscale:** The UHI impact study ranges in a zone 10^2–10^4 m above the city surface. It represents the integrated response of an array of roughness elements with spatial variability reflecting the unique characteristics of different neighborhoods/land uses.
3. **Microscale:** The UHI impact study ranges in a zone 10^1–10^2 m and involves spatial differences in response to individual roughness elements (variability in building/canyon dimensions, trees) and proximity to localised emissions sources (e.g. roads, vegetation).

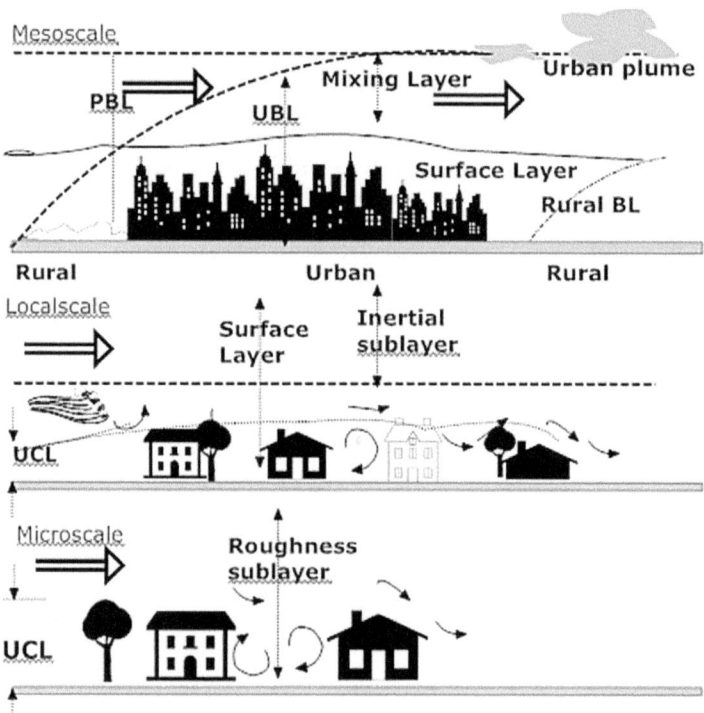

Figure 9.1
Spatial scales
representation
Source: Adapted from [7]
Voogt *et al.*, 1997

The most important ones are the meso- and microclimatic conditions, which differ significantly from that of rural areas. The main reason for this is the alteration of the surface structure (e.g. proportion of the built-up area, 3D geometry of the buildings and trees) triggering particular urban climate phenomena (e.g. urban heat island, changes in the radiation fluxes).

The climate of street canyons is primarily controlled by the micro-meteorological effects of urban geometry rather than the mesoscale forces controlling the climate of the boundary layer. There are strong microscale variations of surface temperature that arise due to changes in radiant load with surface slope and aspect, shading, and variations in surface thermal and radiative properties. [7] Although there are no clear-cut distinctions between different categories, models might be classified into groups according to their physical or mathematical principles (e.g. reduced-scale, box, Gaussian, CFD) and their level of complexity (e.g. screening, semi-empirical, numerical). [8]

Computational Fluid Dynamic (CFD) models

CFD is a numerical method of modelling complex fluid flow by breaking down geometry into cells that comprise a mesh. At each cell, an algorithm is applied to compute the fluid flow for the individual cell. Depending on the nature of the flow, either the Euler or Navier-Stokes equations can be used for the computation.

The CFD software simulation approaches generally followed three steps:

1. *Pre-processing*, **where the urban domain is described using geometry features, CAD files, land-use cover maps, meteorological maps, etc.** CAD software or built-in geometric modelers that are included in various CFD programs can be used to create the building as a geometric representation. After finalising the geometry, a grid/mesh will be generated (Figure 9.2) using a standalone meshing software or built-in meshing tools within the CFD package (such as ANSYS Mesh). The grids are generated using structured Cartesian grids, structured body-fitted grids or unstructured grids. Initial boundary conditions of the urban area are set such as the roughness coefficient and the physical fluid properties are specified.

2. *Solving*, **where a numerical model is defined and the algebraic equations are set for the definition of the flow conditions.** There are four major turbulence

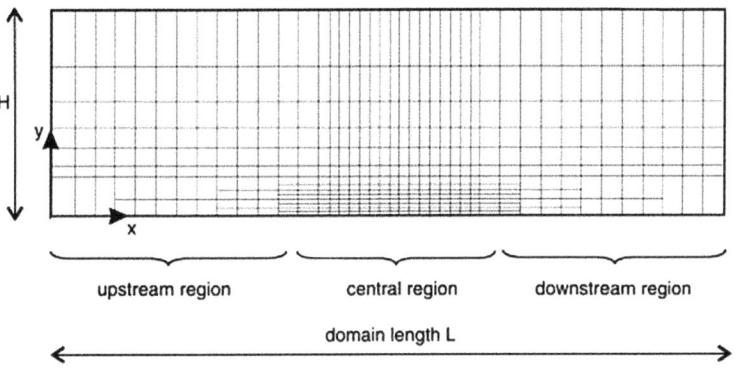

upstream region   central region   downstream region

domain length L

Figure 9.2
Layout of a computational grid for a CFD simulation. Close to the surface area of the central urban domain a higher-density grid is required for a sufficient flow simulation
Source: [9] Blocken *et al.*, 2007. Elsevier Limited, Licence Nr 3593130147523

modelling approaches in CFD, including *Reynolds Averaged Navier Stokes* (RANS), *Large Eddy Simulation* (LES), *Detached Eddy Simulation* (DES) and *Direct Numerical Simulation* (DNS). [10]

The *Reynolds Averaged Navier Stokes* (RANS) approach is a modelling method in which the major variables such as temperature and velocity are divided into time-averaged and turbulent fluctuation, and then the turbulent model is used to model the turbulent fluctuation. Considering that the vertical profiles of the wind velocity or the turbulent kinetic energy within a street canyon are very difficult to determine with on-site measurements for the entire study region grid, these profiles and k-ε turbulence models are very commonly used in all CFD simulation programs (Figure 9.3, Figure 9.4).

Vertical profiles for the wind speed and direction, the temperature and the turbulence, are presented in a very detailed study undertaken under the auspices of COST Action 710, which was concerned with the preprocessing of meteorological data for atmospheric dispersion models [11]. Richards and Hoxey (1993) modelled a horizontally-homogeneous turbulent boundary layer (HHTBL) by proposing velocity and turbulence property profiles, together with the associated boundary conditions, for the standard k-ε turbulence model and showing that these satisfied horizontal homogeneity provided the model constants satisfied particular relationships [12].

$$U(z) = \frac{u_* \ln(z/z_0)}{\sqrt{(c_{\varepsilon 2} - c_{\varepsilon 1})\sigma_\varepsilon \sqrt{c_\mu}}}, \ \ \kappa(z) = \frac{u_*^2}{\sqrt{c_\mu}}, \ \ \varepsilon(z) = \frac{u_*^2}{z\sqrt{(c_{\varepsilon 2} - c_{\varepsilon 1})\sigma_\varepsilon \sqrt{c_\mu}}}$$

where

z is the height coordinate of the grid calculations,

u_* is the friction velocity associated with the constant shear stress $\tau_{xz} = \rho u_*^2$

κ the turbulent kinetic energy

and taking into account the von Karman constant

$$\kappa = \sqrt{(c_{\varepsilon 2} - c_{\varepsilon 1})\sigma_\varepsilon \sqrt{c_\mu}}$$

and

$$\varepsilon(z) = \frac{u_*^2}{z\kappa}, \ \ U(z) = \frac{u_* \ln(z/z_0)}{\kappa}$$

we assume the κ-ε turbulence model, with κ ~ 0.4–0.42, for the corresponding model constants , (i.e. $C_{\varepsilon 1} = 1.44$, $C_{\varepsilon 2} = 1.92$, $C_\mu = 0.09$ and $\sigma_\varepsilon = 1.3$ give the von Karman constant

$$\kappa = \sqrt{(c_{\varepsilon 2} - c_{\varepsilon 1})\sigma_\varepsilon \sqrt{c_\mu}} = 0.4237$$

In *Large Eddie Scale* (LES) models, large-scale structures in turbulent flow are created directly. This is due to unsteadiness of the mean flow (shear or buoyancy effects) that are simulated directly as they are problem reliant, anisotropic and play a key part in transportation of mass, momentum and energy. LES based

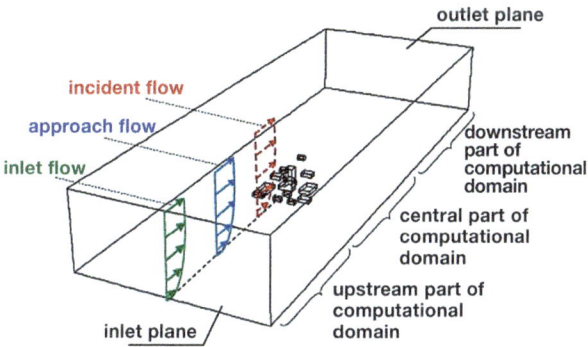

Figure 9.3 Computational domain with building models for CFD simulation of ABL flow –
definition of inlet flow, approach flow and incident flow and indication of
different parts in the domain for roughness modelling
Source: [9] Blocken *et al.*, 2007. Elsevier Limited, Licence Nr 3594650971293

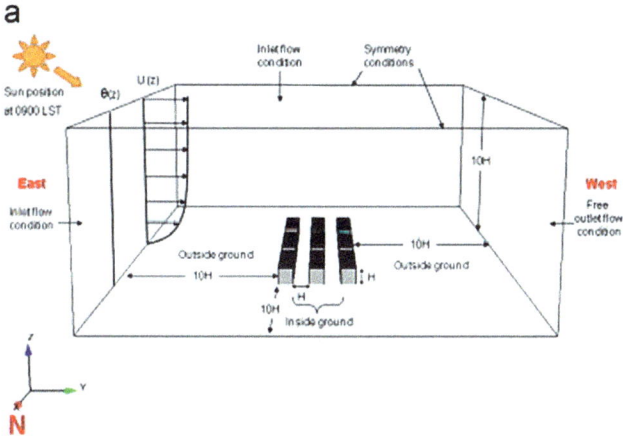

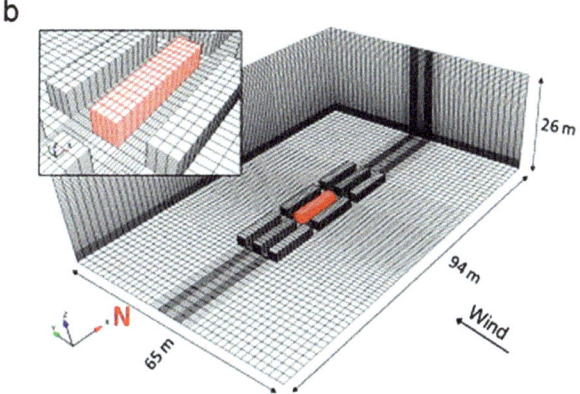

Figure 9.4 Computational domain with building models for low wind speed:
(a) simulations-definition of boundary conditions; (b) mesh of the domain
and sub-domain
Source: [13] Qu *et al.*, (2012). Elsevier Limited, Licence Nr 3595311062446

models provide a far better physical basis for predicting dispersion from accidental releases in urban areas (puff dispersion) or simulating the gustiness of winds in complex urban geometries but, these models need to be validated, considering their intended use.

The introduction of *Detached Eddy Simulation* (DES) was prompted by the quest to address the challenge of high Reynolds number, immensely separated flows.

On the other hand, Direct Numerical Simulation (DNS) is the most accurate technique of solving fluid turbulence. This approach to turbulence simulations solves the Navier-Stokes equations without any approximation of the turbulence except numerical discretisation.

When modelling a 2D urban street canyon, the steady Reynolds-Averaged Navier–Stokes (RANS) approach in combination with the realisable k-ε turbulence model has been shown to predict the mean flow pattern and the transport of passive scalars reasonably well. [14]

According to Z. Xie and I. P. Castro (2006) [15], above the building canopy, steady RANS models results can in general be compared to LES model simulations. Nevertheless, within the canopy, the results obtained from steady RANS models are much less satisfactory compared with LES, which suggests that the inherent unsteadiness there plays an important role in such flows. It should also be stressed that a pure unsteady RANS method could not successfully be applied for such flows in view of the lack both of any significant scale separation and of any significant periodicity. Although the improvement from LES for such flows is at the cost of much more CPU time, e.g. at least an order of magnitude more than RANS, with the rapid advances in computing hardware and software, such an expense is increasingly becoming affordable even for industrial-scale problems. In spite of its deficiencies, steady RANS modelling with the k–ε model

Figure 9.5
Vertical temperature distribution. (a) With solar/radiation 8 am. (b) With solar/radiation 12 pm (c) With solar/radiation 4 pm. (d) With solar/radiation 0 am. (e) Without solar/radiation 8 am. (f) Without solar/radiation 12 pm. (g) Without solar/radiation 4 pm. (h) Without solar/radiation 0 am.
Source: [17] Kaoru *et al.*, 2011. Elsevier Limited, Licence Nr 3595311291562

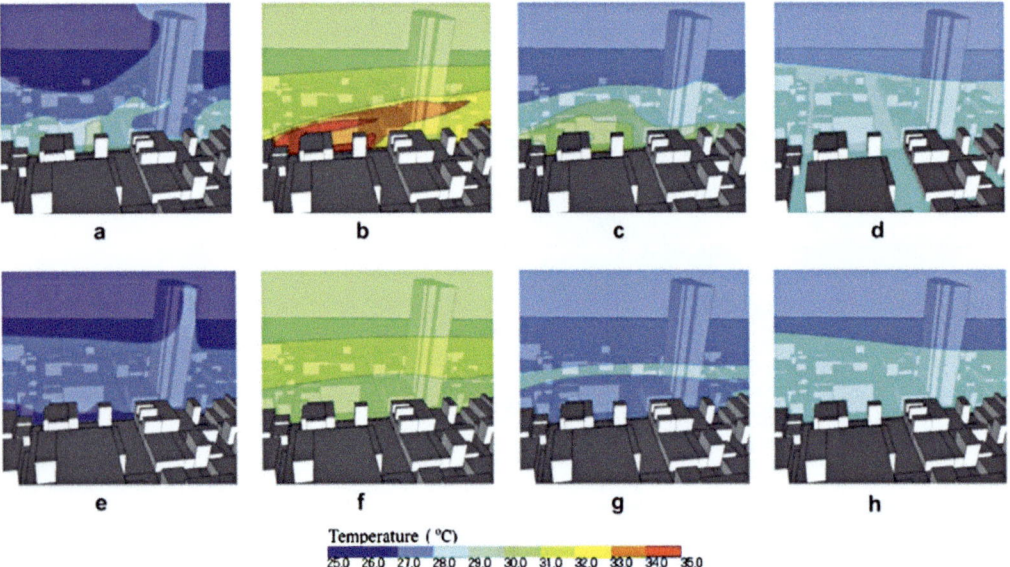

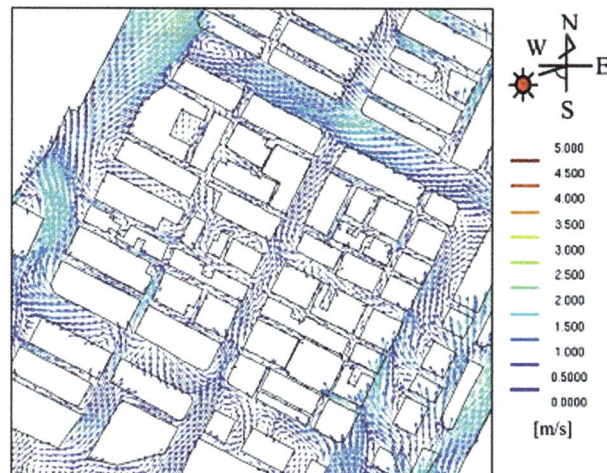

Figure 9.6
Horizontal distribution of the wind velocity in Kyobashi (height: 1.5 m)
Source: [18] Chen *et al.*, 2009. Elsevier Limited, Licence Nr 3595320047810

5.000
4.500
4.000
3.500
3.000
2.500
2.000
1.500
1.000
0.5000
0.0000
[m/s]

or with other turbulence models has become the most popular CFD approach for pedestrian level wind studies. [16].

3. ***Post-processing* includes the presentation of the simulation results**. After calculating the required environmental parameters of each CFD grid point, an adequate layout is included in each software, which presents the results in a graphical user interface (GUI) with visualised layers and maps (Figure 9.5, Figure 9.6).

CFD can determine to a possible degree the urban wind flow and heat transfer. Contemporary UHI research gives weight to analyse complex cases aiming at such accuracy, which will allow the evaluation of the effectiveness of adaptation and mitigation measures in real urban areas. Therefore, the most accurate scale for analysing the urban microclimate is the microscale, in real urban areas, followed by the appropriate validation with real on-site measurements. The simulations should consider as many as possible of the defined causes for the UHI and should follow an unsteady approach. [19]

The possible UHI causes identified in the microscale urban environment are listed below, [19], [20]:

1. *Amplified short-wave radiation gain*. This means that more solar radiation is absorbed because of multiple reflections in the urban environment. In a built environment, solar energy is absorbed into infrastructure paving, roofs and facades of different materials, causing the increase of the surface tempera-ture of urban structures to become 10–20°C higher than the ambient air temperatures [21].
2. *Amplified long-wave radiation gain from the sky*. The increased air pollution over urban areas increases the radiation load.
3. *Decreased long-wave radiation loss*. The radiation losses are decreased and 'trapped' within the street canyons because of the high buildings obstacles.
4. *Anthropogenic heat sources inside urban areas*, resulting from the human activity, the cars or industry emissions, etc.
5. *Increased heat storage*, which arises from the physical properties of con-struction materials (pavement, asphalt, building facades, etc.).

6. *Less evapotranspiration.* Urban areas are covered with less vegetation or water areas, which leads to less moisture availability in the urbanized regions.
7. *Decreased turbulent heat transport.* The city architectural layout with the street canyons can lower the wind velocity and result a reduced convective heat transfer and the limitation of the urban ventilation.

Advantages of CFD modelling for UHI mitigation techniques

When assessing the urban environmental conditions regarding the UHI mitigation, CFD models can compromise significant advantage factors, such as time, cost and ease of visualising results. CFD can be used to evaluate a range of issues comprising air speed and movement, air quality and pollution diffusion, wind or thermal comfort, as well as the effects of different façade, roofing and paving materials or the effects of increasing vegetation and water areas ([38],

Table 9.1 Overview of CFD studies on thermal environment in urban areas

Study (Ref)	Scale	Configuration	Causes considered*	Steady vs. unsteady	Validation
Mochida et al. (1997) [22]	Mesoscale (3D)	No buildings	1, 2, 3, 4, 5, 6	Unsteady	Yes
Bruse and Fleer (1998) [23]	Microscale (3D)	3D generic urban	1, 3, 5, 6, 7	Unsteady	No
Ashie et al. (1999) [24]	Microscale (2D)	2D street canyon	1, 4, 6, 7	Unsteady	No
Ca et al. (1999) [25]	Microscale (2D)	2D buildings	1, 3, 4, 6	Unsteady	Yes
Fujino et al. (1999) [26]	Mesoscale (3D)	No buildings	1, 2, 5, 6	Unsteady	Yes
Takahashi et al. (2004) [27]	Microscale (3D)	3D applied urban	1, 2, 3, 6, 7	Unsteady	Yes
Chen et al. (2004) [28]	Microscale (3D)	3D applied urban	1, 3, 4, 6, 7	Unsteady	Yes
Li et al. (2005) [29]	Microscale (3D)	Single building	1, 2, 3, 5	Quasi-steady	Yes
Murakami (2006) [30]	Microscale (3D)	3D generic urban	1, 3, 4, 6, 7	Unsteady	No
Priyadarsini et al. (2008) [31]	Microscale (3D)	3D applied urban	1, 4, 5, 7	Unsteady	Yes
Chen et al. (2008) [32]	Microscale (3D)	3D generic urban	1, 3, 7	Unsteady	No
Lin et al. (2008) [33]	Microscale (3D)	3D generic urban	1, 2, 5, 6, 7	Unsteady	No
Dimitrova et al. (2009) [34]	Microscale (3D)	3D generic urban	1, 2, 7	Steady	Yes
Chen et al. (2009) [18]	Microscale (3D)	3D applied urban	1, 3, 4, 5, 6, 7	Unsteady	No
Hsieh et al. (2010) [35]	Microscale (3D)	3D applied urban	1, 2, 5, 6, 7	Unsteady	No
Memon et al. (201) [36]	Microscale (2D)	2D street canyon	1, 7	Steady	Yes
Memon and Leung (2010) [37]	Microscale (2D)	2D street canyon	7	Steady	Yes
Kaoru et al. (2011) [17]	Microscale (3D)	3D applied urban	1, 2, 3, 5, 7	Unsteady	No
Berkovic et al. (2012) [38]	Microscale (3D)	3D generic courtyard	1, 2, 3, 5, 6, 7	Unsteady	Yes
Ma et al. (2012) [39]	Microscale (3D)	3D applied urban	1, 2, 5, 6, 7	Unsteady	Yes
Bo-ot et al. (2012) [40]	Microscale (3D)	3D generic urban	3, 4, 7	Unsteady	No
Qu et al. (2012) [13]	Microscale (3D)	3D generic urban	1, 2, 3, 5, 7	Unsteady	No
Yang et al. (2013) [3]	Microscale (3D)	3D applied urban	1, 2, 3, 5, 6, 7	Unsteady	Yes
Allegrini et al. (2014) [41]	Microscale (2D)	2D street canyon	7	Steady	Yes
Toparlar et al. (2014) [19]	Microscale (3D)	3D applied urban	1, 2, 3, 5, 6, 7	Unsteady	Yes
Santamouris et al. (2012) [42]	Microscale (3D)	3D applied urban	5	Steady	No
Maragkogiannis et al. [43] (2014)	Microscale (3D)	3D applied urban	1, 2, 3, 5, 6, 7	Steady	No

*(1) amplified short-wave radiation gain; (2) amplified long-wave radiation gain from the sky; (3) decreased long-wave radiation loss; (4) anthropogenic heat sources; (5) increased heat storage; (6) less evapotranspiration; and (7) decreased turbulent heat transport.

[44], [31], [40], [42], [45]). CFD modelling is more cost effective than on-site assessments and measurements. It can be applied in full urban scales and is not restricted by a wind tunnels' cross-section area. Moreover, it can calculate the wind velocity and temperature values at all the grid points of the urban model, not only at measuring points. This results to a significant acceleration during the urban design development, which facilitates engineers to consider many options of buildings or open spaces taking into account the outdoor thermal comfort conditions.

However, CFD requires a great computational potential depending on the modelled regions' area and the building density, which can often hinder the wind and temperature flow calculations. In such cases, on-site temperature and wind flow measurements can be coupled with the CFD model. Furthermore, the microscale CFD modelling requires a strong user's background on fluid dynamics. The user shall always be in position to explain the model's results or identify irrational results and non-apparent faults.

PHOENICS (Parabolic, Hyperbolic Or Elliptic Numerical Integration Code Series)

PHOENICS was developed by CHAM (Concentration Heat and Momentum Limited) and is a multi-purpose Computational Fluid Dynamics (CFD) software package. [46]

It simulates a range of processes involving fluid flow, heat or mass transfer, chemical reaction and/or combustion in engineering equipment and the environment.

A wide range of environmental conditions can be assessed in a fraction of the time and at a fraction of the cost to carry out a field experiment or wind-tunnel test. Regarding the UHI assessment it can simulate and calculate a variety of physical and structural characteristics such as:

- Convection, conduction and radiation; conjugate heat transfer, with a library of solid materials and automatic linkage at the solid fluid interface.
- A wide range of built-in turbulence models for high and low Reynolds number flows including:

  1. LVEL model for turbulence in congested domains. The LVEL turbulence model is a unique feature of PHOENICS, which is especially useful for conjugate-heat-transfer problems.
  2. k-ε models including RNG, two-scale and two-layer models.

- Multi-phase flows of three kinds with a variety of built-in interphase-transfer models:

  1. Inter-penetrating continua, including turbulence and modulation;
  2. Particle tracking, including turbulence dispersion effects;
  3. Free-surface flows.

- Advanced radiation models, including surface-surface model with calculated view factors, a six-flux model and composite radiosity model for radiative heat transfer, known as IMMERSOL.
- Mechanical and thermal stresses in immersed solids can be computed at the same time as the fluid flow and heat transfer.

Figure 9.7
Representation of the wind velocity simulation results in a central square area in Chania – Crete.
Source: [43] Maragkogiannis *et al.*, 2014. Elsevier Limited, Licence Nr 3593120522708

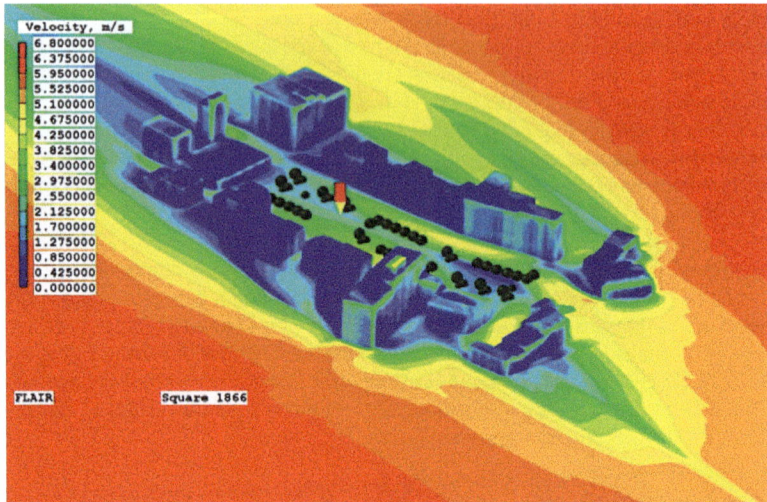

Regarding the PHOENICS UHI applications [47], the CFD analysis with PHOENICS provides very detailed information regarding the climatic conditions of the outdoor space such as airflow velocities and air temperatures as well as the solar absorptivity of the building surfaces. The optical and thermal characteristics of the materials and buildings, the solar radiation pattern as well as the wind flow can be modelled. Moreover, urban thermal environment improvement scenarios can be analysed. [43]

ANSYS FLUENT

FLUENT is a Flow Modelling Software owned and distributed by ANSYS. It is used to model fluid flow within a defined geometry using the principles of computational fluid dynamics. It is available with GAMBIT, a tool to generate or import geometry so that it can be used as a basis for simulations run in FLUENT. It can model various scenarios using computational fluid dynamics, including compressible and incompressible flow, multiphase flow, combustion, and heat transfer.

As it is described in recent reviews [48], FLUENT software is successfully used in order to predict the temperature distribution inside earth-to-air heat exchangers, and to study the passive cooling dissipation techniques for buildings. The accuracy of the proposed method to calculate convective heat transfer is validated against data collected in an experimental chamber. Earth-to-air heat exchangers have been applied in various types of buildings.

OpenFOAM

OpenFOAM is a turbulence model that offers a large range of methods and models to simulate turbulence. [49] OpenFOAM is developed on Linux/UNIX systems and written in C++. However, it is also made operable under Cygwin/X, which is a port in the X Windows System of Microsoft Windows. Like other CFD codes, OpenFOAM simulation contains three processes: pre-processing, solving and post-processing. [50]

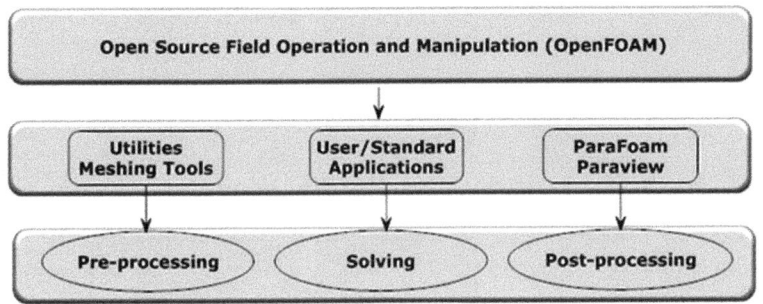

Figure 9.8
OpenFoam Structure
representation
Source: Author generated
from [50] Huang & Wang
2013

The methods include:

- Reynolds-Averaged Simulation (RAS), also known as Reynolds-Averaged Navier-Stokes (RANS): The governing equations are solved in ensemble-averaged form, including appropriate models for the effect of turbulence. See the list of incompressible and compressible RAS models for more information.
- Large Eddy Simulation (LES): Large turbulent structures in the flow are resolved by the governing equations, while the effect of the sub-grid scales (SGS) are modelled. The scale separation is obtained by applying a filter to the governing equations which also influences the form of the SGS models. The list of LES models contains information about models, LES filters and filter width functions.
- Detached Eddy Simulation (DES): Hybrid method that treats near-wall regions with a RAS approach and the bulk flow with an LES approach. DES models are found in the list of LES models.
- Direct Numerical Simulation (DNS): Resolves all scales of turbulence by solving the Navier-Stokes equations numerically without any turbulence modelling. DNS simulations can be performed in OpenFOAM using the dnsFoam solver.

Urban thermal environment models

ENVI-met

ENVI-met is a three-dimensional microclimate model designed to simulate the surface-plant-air interactions in urban environment with a typical resolution of 0.5 to 10 m in space and 10 sec in time which simulates even complicated geometric forms such as terraces, balconies or complex quarters. It is also a prognostic model based on the fundamental laws of fluid dynamics and thermo-dynamics. Its high spatial resolution allows a fine analysis of the microclimate at street level and the possibility of representing complex geometries including galleries and horizontal overhangs as well as various vegetation covers. [5] Typical areas of application are Urban Climatology, Architecture, Building Design or Environmental Planning. ENVI-met is used to calculate the microclimate and air quality in urban structure and open space. [6] ENVI-met model is capable of predicting the microclimate changes within urban environments in terms of different variables with good accuracy. [1]

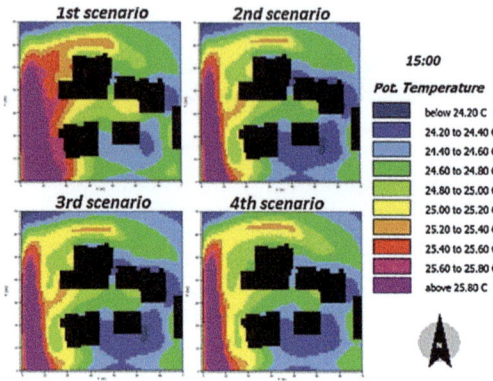

Figure 9.9
Representation of an urban garden case study in ENVIMET, using four scenarios (1st – absence of vegetation, 2nd current situation, 3rd & 4th different types of urban gardens)
Source: [51] Tsilini *et al.*, 2015. Elsevier Limited, Licence Nr 3593121062972

The model includes the simulation of:

1. Flow around and between buildings
2. Exchange processes of heat and vapour at the ground surface and at walls
3. Turbulence
4. Exchange at vegetation and vegetation parameters
5. Bioclimatology
6. Pollutant dispersion

The simulation of the abovementioned parameters is grouped in four main systems:

1. Soil
2. Vegetation
3. Atmosphere
4. Building

The soil from the surface to 2 m depth is divided into multiple layers with the vertical spacings. Different soil profiles and patches of surfaces can be modelled by allocating various natural soils or artificial materials for each grid cell. For natural soils, the heat and water vapor transfers are taken into account while merely heat transfer is considered for sealed materials. The ground surface temperature is calculated from the net radiative fluxes, the turbulent fluxes of heat and vapor and the heat conduction into the ground at the surface. The albedo of natural soil is determined by the model itself, as a function of solar incident angle and water content of topsoil layer.

ENVI-met is a freeware program (not open source) based on different scientific research projects and is therefore under constant development. It comes along with a number of additional software ranging from editors up to graphical visualisation tools for the model results. The model combines the calculation of fluid dynamics parameters such as wind flow or turbulence with the thermodynamic processes taking place at the ground surface, at walls and roofs or at plants.

An inaccuracy consists of the fact that its high complexity makes the model very slow. In order to succeed sufficient results in a specific time period, one has to either limit the calculated timespan or the model resolution or the size of the area of interest. However the limitation to the resolution leads to inaccuracy. Due to the low horizontal grids resolution, all objects become cuboids. While larger objects are composed by several cuboids, smaller objects can't be included in the calculations at all. Furthermore, the low horizontal resolution leads to inaccuracy in the calculation of radiation and air flow. [52]

RayMan model

One of the recently used radiation and bioclimatic models is RayMan, developed in the Meteorological Institute, University of Freiburg. It is well suited to calculate radiation fluxes.

The RayMan model was developed according to guidelines of the German Engineering Society. It calculates the radiation flux within urban structures on the basis of parameters such as air temperature, air humidity, degree of cloud cover, time of day and year, albedo of the surrounding surfaces and their solid-angle proportions.

The main advantage of the RayMan is that it facilitates the reliable determination of the microclimatological modifications of different urban environments, since the model considers the radiation modification effects of the complex surface structure (buildings, trees) very precisely. Beside the meteorological parameters the model requires input data on surface morphological conditions of the study area and on personal parameters.

Mesoscale models

With respect to the thermal impact induced by human activities, the anthropogenic heat flux in the city is a factor that is often ignored in atmospheric models but has recently been shown to be important in the development of the UHI.

The possible UHI causes identified in the mesoscale urban environment are listed below [20]:

1. *Increased absorption of short-wave radiation* (Air pollution – increased aerosol absorption)
2. *Anthropogenic heat source* (traffic/vehicle, building/chimney emissions)
3. *Increased sensible heat input-entrainment from below* (increased heat flux from surface layer and roofs)
4. *Increased sensible heat input entrainment from above* (increased heat mass entrainment)

Anthropogenic heating is the combination of waste heat released from vehicle fuel combustion, building and industrial energy consumption, and human metabolism. The heat intensity varies with climate, population density, and intensity of industrial and commercial activities. While the average anthropogenic heat flux is small compared with summertime midday solar radiation, it may play a major role in the urban energy budget during the night and winter, when the urban heat island effect is most prominent and the relative solar forcing is low. [53]

MM5 is a three-dimensional prognostic model that is widely used in numerical weather prediction and air quality studies. Since energy consumption processes occur at or near the urban surface, and hence within the atmospheric boundary layer, the effect of anthropogenic heating is incorporated into the subroutines that model PBL turbulent mixing processes. Since these turbulent fluxes of heat, moisture and momentum cannot be explicitly resolved by mesoscale models, they are parameterised via a PBL scheme. There have been various PBL schemes developed for mesoscale modelling. For example, the MM5 itself provides seven options for modelling PBL processes. Each available PBL scheme is unique, but can be categorised into one of three general turbulent mixing approaches: (1) local first-order closure models; (2) non-local exchange models; and (3) high-order closure models. The latter two approaches are commonly used in mesoscale modelling,

Pennsylvania State University/National Center for Atmospheric Research mesoscale model MM5

The MM5 Model is a Mesoscale Modelling system designed to simulate and predict mesoscale atmospheric circulation. MM5 model (the fifth generation NCAR / Penn State Mesoscale Model) has been widely used for tropical storms and other important mesoscale weather process simulation, and environmental science. [54], [55]

Its features include:

1. a multiple-nest capability
2. non-hydrostatic dynamics, which allows the model to be used at a few-kilometer scale
3. multitasking capability on shared- and distributed-memory machines
4. a four-dimensional data-assimilation capability, and
5. several physics options.

The model is a public domain programme.

Numerical simulations can be performed and supplemented by information derived from satellite image analysis for greater urban areas. [44]

Weather Research Forecasting (WRF) model

The Weather Research and Forecasting (WRF) is an atmospheric model developed for both research and operational applications and can be used for simulations across varying spatial scales from few kilometer to hundreds of kilometer. The model was developed as a collaborative effort among various institutes like the National Centre for Atmospheric Research (NCAR), the National Centers for Environmental Prediction (NCEP), the National Oceanic and Atmospheric Administration (NOAA), the Forecast Systems Laboratory (FSL) and the Air Force Weather Agency (AFWA). [56]

The WRF model (Skamarock et al., 2005) is a non-hydrostatic, compressible model with a mass coordinate system. It was designed as a numerical weather prediction model, but can also be applied as a regional climate model. It has a number of options for various physical processes. [57] At the larger scales of the mesoscale modelling, numerical weather prediction (NWP) models incorporate sophisticated urban canopy model (UCM) parameterisations that account for the effects of the urban canopy, including the increased mechanical drag,

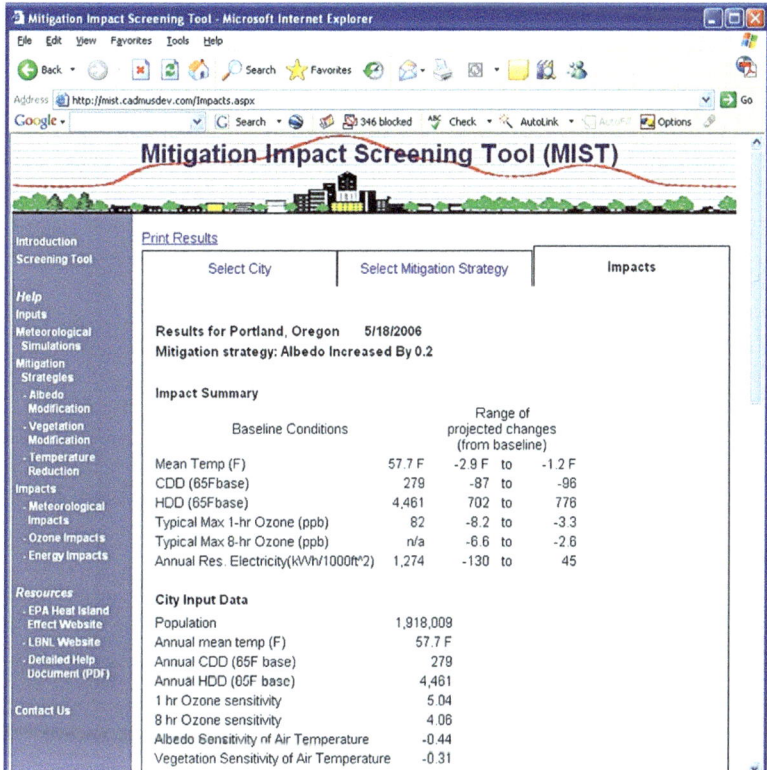

Figure 9.10
MIST impact summary for a
mitigation scenario in
Oregon, Portland, 2006.

turbulence production, and the thermal effects leading to formation of urban heat islands. These parameterisations only consider the average properties of the urban canopy structures, including the mean building heights, the alignment of the street canyons compared to the mean wind speed and the thermal properties of buildings and street canyons. [58]

Mitigation Impact Screening Tool (MIST)

The Mitigation Impact Screening Tool (MIST) is web-based software for the estimation of the UHI mitigation strategies potential regarding the urban climate, the air quality and the energy consumption within the cities. It combines the necessity for a cost-effective and prompt assessment of mitigation strategies in a wide range of cities and the requirement that the results from the tool be based on sound scientific methods

The general approach behind MIST involved first developing detailed meteorological model simulations for a set of twenty test cities in the USA.

For the analysis, the MM5 mesoscale model from the National Center for Atmospheric Research was used. The PSU/NCAR mesoscale model (known as MM5) is a limited-area, non-hydrostatic, terrain-following sigma-coordinate model designed to simulate or predict mesoscale atmospheric circulation. [55]

Figure 9.11
Urban climate zone
classification
Source: [62] Oke 2006

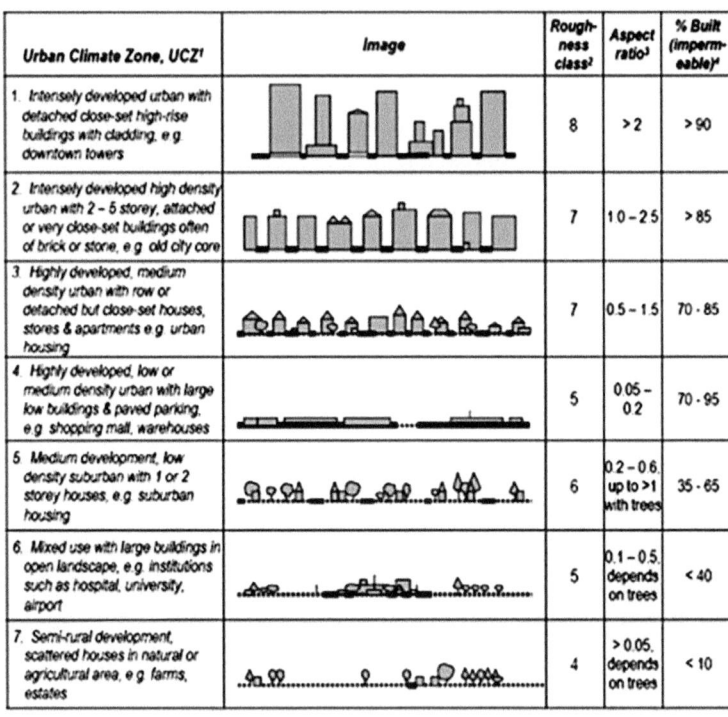

Urban Climate Zone, UCZ[1]	Image	Rough-ness class[2]	Aspect ratio[3]	% Built (imperm-eable)[4]
1. Intensely developed urban with detached close-set high-rise buildings with cladding, e.g. downtown towers		8	>2	>90
2. Intensely developed high density urban with 2 – 5 storey, attached or very close-set buildings often of brick or stone, e.g. old city core		7	1.0 – 2.5	>85
3. Highly developed, medium density urban with row or detached but close-set houses, stores & apartments e.g. urban housing		7	0.5 – 1.5	70 - 85
4. Highly developed, low or medium density urban with large low buildings & paved parking, e.g. shopping mall, warehouses		5	0.05 – 0.2	70 - 95
5. Medium development, low density suburban with 1 or 2 storey houses, e.g. suburban housing		6	0.2 – 0.6, up to >1 with trees	35 - 65
6. Mixed use with large buildings in open landscape, e.g. institutions such as hospital, university, airport		5	0.1 – 0.5, depends on trees	< 40
7. Semi-rural development, scattered houses in natural or agricultural area, e.g. farms, estates		4	> 0.05, depends on trees	< 10

Key to image symbols: ▮ buildings; ♀ vegetation, ▬▬ impervious ground; ∙∙∙∙∙∙ pervious ground

[1] A simplified set of classes that includes aspects of the schemes of Auer (1978) and Ellefsen (1990/91) plus physical measures relating to wind, thermal and moisture controls (columns at right). Approximate correspondence between UCZ and Ellefsen's urban terrain zones is: 1(Dc1, Dc8), 2 (A1-A4, Dc2), 3 (A5, Dc3-5, Do2), 4 (Do1, Do4, Do5), 5 (Do3), 6 (Do6), 7 (none).
[2] Effective terrain roughness according to the Davenport classification (Davenport et al. 2000); see Table 2.
[3] Aspect ratio = z_H/W is average height of the main roughness elements (buildings, trees) divided by their average spacing, in the city centre this is the street canyon height/width. This measure is known to be related to flow regime types (Oke 1987) and thermal controls (solar shading and longwave screening) (Oke, 1981). Tall trees increase this measure significantly.
[4] Average proportion of ground plan covered by built features (buildings, roads, paved and other impervious areas) the rest of the area is occupied by pervious cover (green space, water and other natural surfaces). Permeability affects the moisture status of the ground and hence humidification and evaporative cooling potential.

Local climate zones

Besides the recent decades computational software evolution for the urban climate assessment, there have been also developed empirical methods for the representation of the urban scale and for the classification of the urban or rural measurement sites defining heat island magnitude [59]. For different purposes, scientists have developed landscape classification adapted to urban areas and oriented for climate studies. Urban climate maps can be considered as the first spatially exhaustive and expert-knowledge approach to provide information and planning recommendations that integrate climate factors. [60] As a result, the "local climate zone" (LCZ) classification system has provided a new research basis for urban heat island studies and has regulated in a worldwide scale the exchange of urban temperature observations. [61] In this empirical method terrestrial data, surface maps (land use /land cover) and analytical climate maps (air temperature, atmospheric humidity, wind direction.) are combined, forming *urban climate maps*. Those studies can be enriched with urban morphological information such as building volume, ground coverage ratio, blue or green spaces, etc.). T. R. Oke's UCZ model (2004) [62] is the most representative since it is included in the World Meteorological Organisation guidelines. This classification has defined the Urban Climate Zones (UCZ) (Figure 9.11) from theoretical divisions of the urban surface based on the local climate alterations. This graphic exemplarily divides urban areas into seven discrete, homogenous units, or "zones," defined only by their ability (in terms of surface geometry and cover) to modify the local surface climate. Each of the seven UCZs has a characteristic roughness class, aspect ratio, and coverage percentage. This ability is in sequence connected with the surface's properties like built fraction, roughness class, and aspect ratio, which inherently co-vary with urban landscape structure, land cover, building fabric, and metabolism (anthropogenic heat, water and pollution). These zones can be used at the local scale and as a general guide for the siting and exposure of urban climate stations.

Nevertheless, although there is a current need on research to link surface temperatures and physical properties of the urban landscape, and also the need for urban climatic maps that can be used by city planners or public authorities, climate measurements for UHI monitoring are always of great importance, for the scientific observations of the present and future urban's climate and hence life quality.

References

[1]   M. K. Svensson and I. Eliasson, "Diurnal air temperatures in built-up areas in relation to urban planning," *Landsc. Urban Plan.*, vol. 61, no. 1, pp. 37–54, Sep. 2002.

[2]   I. N. Harman, "The energy balance of urban areas," (Doctoral thesis, University of Reading). Oct. 2003. [Online]. Available: http://www.met.rdg.ac.uk/phdtheses/The %20energy%20balance%20of%20urban%20areas.pdf [Accessed November 2014].

[3]   X. Yang, L. Zhao, M. Bruse, and Q. Meng, "Evaluation of a microclimate model for predicting the thermal behavior of different ground surfaces," *Build. Environ.*, vol. 60, pp. 93–104, Feb. 2013.

[4]   J. F. Barlow, "Progress in observing and modelling the urban boundary layer," *Urban Clim.*, vol. 10, pp. 216–240, Aug. 2014.

[5]   I. Eliasson, "The use of climate knowledge in urban planning," *Landscape and Urban Planning*, vol. 48, pp. 31–44, 2000.

[6]   D. L. Meteorology and L. Prandtl, "An introduction to the atmospheric boundary layer (ABL)," 2008. [Online]. Available: http://www.met.rdg.ac.uk/~swrhgnrj/teaching/ MT36E/MT36E_BL_lecture_notes.pdf [Accessed: Feb. 7, 2014].

[7]   J. A. Voogt and T. R. Oke, "Complete urban surface temperatures," *Appl. Meteorol.*, vol. 36, pp. 1117–1132, 1997.

[8]   S. Vardoulakis, B. E. A. Fisher, K. Pericleous, and N. Gonzalez-Flesca, "Modelling air quality in street canyons: a review," *Atmos. Environ.*, vol. 37, pp. 155–182, 2003.

[9]   B. Blocken, T. Stathopoulos, and J. Carmeliet, "CFD simulation of the atmospheric boundary layer: Wall function problems," *Atmos. Environ.*, vol. 41, no. 2, pp. 238–252, Jan. 2007.

[10]   N. Hamza, K. Setaih, M. A. Mohammed, S. Dudek, and T. Townshend, "CFD modelling as a tool for assessing outdoor thermal comfort conditions in urban settings in hot arid climates," *Journal of Information Technology in Construction* vol. 19, pp. 248–269, June 2014.

[11]   B. E. A. Fisher, J. J. Erbrink, S. Finardi, P. Jeannet, S. Joffre, M. G. Morselli, U. Pechinger, P. Seibert and D. J. Thomson. Research and Development 1998 Directorate-General Science, "Harmonisation of the pre-processing of meteorological data for atmospheric dispersion models," COST Action 710 – Final report. 1998.

[12] S. E. Norris and P. J. Richards, "Appropriate boundary conditions for computational wind engineering models revisited," Presented at: The Fifth International Symposium on Computational Wind Engineering (CWE2010) Chapel Hill, North Carolina, USA May 23–27, 2010.

[13] Y. Qu, M. Milliez, L. Musson-Genon, and B. Carissimo, "Numerical study of the thermal effects of buildings on low-speed airflow taking into account 3D atmospheric radiation in urban canopy," J. Wind Eng. Ind. Aerodyn., vol. 104–106, pp. 474–483, May 2012.

[14] T. L. Chan, G. Dong, C. W. Leung, C. S. Cheung, and W. T. Hung, "Validation of a two-dimensional pollutant dispersion model in an isolated street canyon," Atmospheric Environment, vol. 36, pp. 861–872, 2002.

[15] Z. Xie and I. P. Castro, "LES and RANS for Turbulent Flow over Arrays of Wall-Mounted Obstacles," Flow, Turbul. Combust., vol. 76, no. 3, pp. 291–312, Aug. 2006.

[16] B. Blocken, "50 years of Computational Wind Engineering: Past, present and future," J. Wind Eng. Ind. Aerodyn., vol. 129, pp. 69–102, 2014.

[17] I. Kaoru, K. Akira, and K. Akikazu, "The 24-h unsteady analysis of air flow and temperature in a real city by high-speed radiation calculation method," Build. Environ., vol. 46, no. 8, pp. 1632–1638, Aug. 2011.

[18] H. Chen, R. Ooka, H. Huang, and T. Tsuchiya, "Study on mitigation measures for outdoor thermal environment on present urban blocks in Tokyo using coupled simulation," Build. Environ., vol. 44, no. 11, pp. 2290–2299, Nov. 2009.

[19] Y. Toparlar, B. Blocken, P. Vos, G. J. F. van Heijst, W. D. Janssen, T. van Hooff, H. Montazeri, and H. J. P. Timmermans, "CFD simulation and validation of urban microclimate: A case study for Bergpolder Zuid, Rotterdam," Build. Environ., vol. 83, pp. 79–90, Aug. 2014.

[20] T. R. Oke, "The energetic basis of the urban heat island," Q. J. R. Meteorol. Soc., vol. 108, no. No. 455, pp. 1–24.

[21] H. Taha, D. Sailor, and H. Akbari. "High-Albedo Materials for Reducing Building Cooling Energy Use," California Inst. for Energy Efficiency, Berkeley, CA, 1992.

[22] H. S. Akashi Mochida, S. Murakami, T. Ojima, S. Kim, R. Ooka. "CFD analysis of mesoscale climate in the Greater Tokyo area," Wind Engineering and Industrial Aerodynamics, vol. 68, pp. 459–477, 1997.

[23] M. Bruse and H. Fleer, "with a three dimensional numerical model," Environmental Modelling & Software, vol. 13, pp. 373–384, 1998.

[24] Y. Ashie, V. T. Ca, and T. Asaeda, "Building canopy model for the analysis of urban climate," Journal of Wind Engineering & Industrial Aerodynamics, vol. 81, 1999. Available: http://www.sciencedirect.com/science/article/pii/S0167610599000203.

[25] V. T. Ca, T. Asaeda, and Y. Ashie, "Development of a numerical model for the evaluation of the urban thermal environment," Journal of Wind Engineering & Industrial Aerodynamics, vol. 81, pp. 181–196, 1999.

[26] T. Fujino, T. Asaeda, and V. T. Ca, "Numerical analyses of urban thermal environment in a basin climate } application of a k } model to complex terrain," Journal of Wind Engineering & Industrial Aerodynamics, vol. 81, pp. 159–169, 1999.

[27] K. Takahashi, H. Yoshida, Y. Tanaka, N. Aotake, and F. Wang, "Measurement of thermal environment in Kyoto city and its prediction by CFD simulation," Energy Build., vol. 36, no. 8, pp. 771–779, Aug. 2004.

[28] H. Chen, R. Ooka, K. Harayama, S. Kato, and X. Li, "Study on outdoor thermal environment of apartment block in Shenzhen, China with coupled simulation of convection, radiation and conduction," Energy Build., vol. 36, no. 12, pp. 1247–1258, Dec. 2004.

[29] X. Li, Z. Yu, B. Zhao, and Y. Li, "Numerical analysis of outdoor thermal environment around buildings," Build. Environ., vol. 40, no. 6, pp. 853–866, Jun. 2005.

[30] S. Murakami, "Environmental design of outdoor climate based on CFD," Fluid Dyn. Res., vol. 38, no. 2–3, pp. 108–126, Feb. 2006.

[31] R. Priyadarsini, W. N. Hien, and C. K. Wai David, "Microclimatic modelling of the urban thermal environment of Singapore to mitigate urban heat island," *Sol. Energy*, vol. 82, no. 8, pp. 727–745, Aug. 2008.

[32] H. Chen, R. Ooka, and S. Kato, "Study on optimum design method for pleasant outdoor thermal environment using genetic algorithms (GA) and coupled simulation of convection, radiation and conduction," *Build. Environ.*, vol. 43, no. 1, pp. 18–30, Jan. 2008.

[33] B. Lin, X. Li, Y. Zhu, and Y. Qin, "Numerical simulation studies of the different vegetation patterns' effects on outdoor pedestrian thermal comfort," *J. Wind Eng. Ind. Aerodyn.*, vol. 96, no. 10–11, pp. 1707–1718, Oct. 2008.

[34] R. Dimitrova, J.-F. Sini, K. Richards, M. Schatzmann, M. Weeks, E. Perez García, and C. Borrego, "Influence of Thermal Effects on the Wind Field Within the Urban Environment," *Boundary-Layer Meteorol.*, vol. 131, no. 2, pp. 223–243, Mar. 2009.

[35] C.-M. Hsieh, H. Chen, R. Ooka, J. Yoon, S. Kato, and K. Miisho, "Simulation analysis of site design and layout planning to mitigate thermal environment of riverside residential development," *Build. Simul.*, vol. 3, no. 1, pp. 51–61, Mar. 2010.

[36] R. A. Memon, D. Y. C. Leung, and C.-H. Liu, "Effects of building aspect ratio and wind speed on air temperatures in urban-like street canyons," *Build. Environ.*, vol. 45, no. 1, pp. 176–188, Jan. 2010.

[37] R. A. Memon and D. Y. C. Leung, "On the heating environment in street canyon," *Environ. Fluid Mech.*, vol. 11, no. 5, pp. 465–480, Dec. 2010.

[38] S. Berkovic, A. Yezioro, and A. Bitan, "Study of thermal comfort in courtyards in a hot arid climate," *Sol. Energy*, vol. 86, no. 5, pp. 1173–1186, May 2012.

[39] J. Ma, X. Li, and Y. Zhu, "A simplified method to predict the outdoor thermal environment in residential district," *Build. Simul.*, vol. 5, no. 2, pp. 157–167, May 2012.

[40] L. M. Bo-ot, Y.-H. Wang, C.-M. Chiang, and C.-M. Lai, "Effects of a Green Space Layout on the Outdoor Thermal Environment at the Neighborhood Level," *Energies*, vol. 5, no. 12, pp. 3723–3735, Sep. 2012.

[41] J. Allegrini, V. Dorer, and J. Carmeliet, "Buoyant flows in street canyons: Validation of CFD simulations with wind tunnel measurements," *Build. Environ.*, vol. 72, pp. 63–74, Feb. 2014.

[42] M. Santamouris, N. Gaitani, A. Spanou, M. Saliari, K. Giannopoulou, and T. Vasilakopoulou, K Kardomateas, "Using cool paving materials to improve micro-climate of urban areas – Design realization and results of the Flisvos project," *Build. Environ.*, vol. 53, pp. 128–136, 2012.

[43] K. Maragkogiannis, D. Kolokotsa, E. Maravelakis, and a. Konstantaras, "Combining terrestrial laser scanning and computational fluid dynamics for the study of the urban thermal environment," *Sustain. Cities Soc.*, vol. 13, pp. 207–216, Oct. 2014.

[44] A. Synnefa, A. Dandou, M. Santamouris, M. Tombrou, and N. Soulakellis, "On the Use of Cool Materials as a Heat Island Mitigation Strategy," *J. Appl. Meteorol. Climatol.*, vol. 47, no. 11, pp. 2846–2856, Nov. 2008.

[45] M. Santamouris, "Cooling the cities – A review of reflective and green roof mitigation technologies to fight heat island and improve comfort in urban environments," *Sol. Energy*, vol. 103, pp. 682–703, May 2014.

[46] Concentration, Heat and Momentum Limited (CHAM), "Cham-Phoenics," 2014. [Online]. Available: http://www.cham.co.uk/.

[47] CHAM, "Prototype Heat Island Application," 2011. [Online]. Available: http://www.cham.co.uk/casestudies/CCS_HeatIsland_Application.pdf [Accessed January 2015].

[48] M. Santamouris and D. Kolokotsa, "Passive cooling dissipation techniques for buildings and other structures: The state of the art," *Energy Build.*, vol. 57, pp. 74–94, Feb. 2013.

[49] OpenFOAM Foundation, "OpenFOAM® – The Open Source Computational Fluid Dynamics (CFD) Toolbox," 2014. [Online]. Available: http://www.openfoam.org/.

[50] Y. Huang & Y. Wang, "Urban Wind and Thermal Environment Simulation – A Case Study of Gävle, Sweden," University of Gaevle, Faculty of Engineering and Sustainable Development, 2013. [Online] Available: http://hig.diva-portal.org/smash/get/diva2:715365/FULLTEXT02.pdf [Accessed February 2015].

[51] V. Tsilini, S. Papantoniou, D.-D. Kolokotsa, and E.-A. Maria, "Urban gardens as a solution to energy poverty and urban heat island," *Sustain. Cities Soc.*, vol. 14, pp. 323–333, Feb. 2015.

[52] D. Fröhlich and A. Matzarakis, "Modelling of changes in human thermal bioclimate resulting from changes in urban design – Example based on a popular place in Freiburg, SW-Germany," *Proc. MettoolsVIII, 8. Fachtagung des Ausschusses Umweltmeteorologie der Dtsch. Meteorol. Gesellschaft. 20.3 bis 22.3.2012, Leipzig. 5-P-4, 1–8*, pp. 1–8, 2012.

[53] H. Fan and D. Sailor, "Modelling the impacts of anthropogenic heating on the urban climate of Philadelphia: a comparison of implementations in two PBL schemes," *Atmos. Environ.*, vol. 39, no. 1, pp. 73–84, Jan. 2005.

[54] G. A. Grell, J. Dudhia and D. R. Stauffer. "A Description of the Fifth-Generation Penn State / NCAR Mesoscale Model (MM5)," no. NCAR/TN-398 + STR. [NCAR Technical Note]. [Online] Available: http://nldr.library.ucar.edu/repository/assets/technotes/TECH-NOTE-000-000-000-214.pdf [Accessed December 2014].

[55] National Center for Atmospheric Research Earth System Laboratory, "MM5 Community Model Homepage," *2014*. [Online]. Available: http://www.mmm.ucar.edu/.

[56] R. Shrivastava, S. K. Dash, R. B. Oza, and M. N. Hegde, "Evaluation of parameterization schemes in the Weather Research and Forecasting (WRF) model: A case study for the Kaiga nuclear power plant site," *Ann. Nucl. Energy*, vol. 75, pp. 693–702, Jan. 2015.

[57] F. Chen, H. Kusaka, R. Bornstein, J. Ching, C. S. B. Grimmond, S. Grossman-Clarke, T. Loridan, K. W. Manning, A. Martilli, S. Miao, D. Sailor, F. P. Salamanca, H. Taha, M. Tewari, X. Wang, A. a. Wyszogrodzki, and C. Zhang, "The integrated WRF/urban modelling system: development, evaluation, and applications to urban environmental problems," *Int. J. Climatol.*, vol. 31, no. 2, pp. 273–288, Feb. 2011.

[58] A. A. Wyszogrodzki, S. Miao, and F. Chen, "Evaluation of the coupling between mesoscale-WRF and LES-EULAG models for simulating fine-scale urban dispersion," *Atmos. Res.*, vol. 118, pp. 324–345, Nov. 2012.

[59] I. D. Stewart, "Landscape representation and the urban-rural dichotomy in empirical urban heat island literature, 1950–2006," *ACTA Climatol. Chorol. Univ. Szeged. Tomus 40–41*, no. 1951, pp. 111–121, 2007.

[60] T. Houet and G. Pigeon, "Mapping urban climate zones and quantifying climate behaviors – an application on Toulouse urban area (France)," *Environ. Pollut.*, vol. 159, no. 8–9, pp. 2180–92, 2011.

[61] I. D. Stewart and T. R. Oke, "Local Climate Zones for Urban Temperature Studies," *Bull. Am. Meteorol. Soc.*, vol. 93, no. 12, pp. 1879–1900, Dec. 2012.

[62] T. R. Oke, "Initial Guidance to Obtain Representative Meteorological Observations at Urban Sites," Instruments and Observing Methods Report No. 81, World Meteorological Organization, 2006. [Online]. Available: https://www.wmo.int/pages/prog/www/IMOP/publications/IOM-81/IOM-81-UrbanMetObs.pdf [Accessed September 2014].

10

URBAN HEAT ISLAND MITIGATION TECHNOLOGIES

Case studies

Denia Kolokotsa

Technical University of Crete – Chania, Greece

Introduction

The environmental quality of urban open spaces has become a fundamental subject for town planners, architects and climatologists. The recent research trends show that the creation of a link between the complicated results of the urban climatology, on one hand, and the town planners, on the other hand, demands more design-oriented outcomes.

By carefully studying the state of the art in the specific sector, the assessment of the microclimate in outdoor spaces, as well as the implementation of various measures for the improvement of the outdoor comfort is performed in various regions worldwide.

The analysis of the various case studies demonstrated that a number of variables need to be taken into account for a comprehensive assessment of the urban heat mitigation technologies for the urban spaces. Those are:

- Climatic conditions;
- Type and extent of the urban space (street, square, neighborhood, district, etc.);
- Type of technologies used;
- Mitigation of the urban heat achieved.

The chapter presents a number of real applications of urban heat island mitigation technologies. The case studies refer to different regions and different climatic zones. A dedicated format was implemented for the presentation of the various case studies included in the chapter. The selection of case studies was based on the following criteria: (a) Geographical Coverage; (b) Type of urban space under study (i.e. district, street, neighborhood, parking, etc. (c) Type of urban heat mitigation technology used.

Flisvos Park – Athens, Greece

Locality and climatic conditions

The Flisvos Urban Park is a coastal area located in the south western part of Athens, Greece (Figure 10.1). It is surrounded by a major traffic axis and the sea of Saronic Golf. The greater Athens region belongs to the Mediterranean climatic

Figure 10.1 The Flisvos Park in Athens Greece
Source: Reprinted from Santamouris, M., Gaitani, N., Spanou, A.,
Saliari, M., Giannopoulou, K., and Vasilakopoulou, K.
Kardomateas, T. (2012). Using cool paving materials to improve
microclimate of urban areas – Design realization and results of
the Flisvos project. *Building and Environment*, 53, 128–136.
doi:10.1016/j. buildenv.2012.01.022 with permission from Elsevier

zone with hot summers, mild winters. Moreover the area is quite dry. The climate of Flisvos Park during summer is determined by the transfer of heat by the northern winds, which are heated travelling over the city and the sea bridge caused by the Saronic Gulf. In parallel the local micro-climate is highly influenced by the proximity to the southern highway axis that presents a very high traffic load. Despite the adjacency to the Saronic Bay, the southern zones of the city present temperatures higher by 1–2 K, in comparison to the northern and northeastern parts.

The urban space

The park was composed of green and paved areas that provided access to the sea. The so called green part was quite sparsely planted with small trees and bushes. Paved areas were made of asphalt, concrete and quite dark paving components. The absorptivity of the used paved surfaces was measured between 0.55 to 0.65, while areas covered with concrete and asphalt presented much higher values, 0.79 and 0.89 respectively (Santamouris *et al.*, 2012).

Urban heat island mitigation technologies

In total, 4,500 m² of cool paving materials have been applied to the area, while extensive measurements combined to numerical modeling have been performed before and after the rehabilitation of the area to identify the mitigation potential of the used cool materials and document the thermal performance of the whole project. The overall refurbishment of the park was started at 2009 and was finalized during the summer of 2010. The first phase was finished during the end of 2009, when all green areas had been planted, while the change of the pavements took place during June and part of the July 2010.

Results

The application of cool pavements is analysed using both monitoring data and computational fluid dynamics. The analysis of the results showed almost 2 K reduction of the air temperature and 5.5–7 K of the surface temperature.

Bioclimatic Boulevard, Vallecas, Madrid Spain

Locality and climatic conditions

In the framework of LIFE ECO-Valley Mediterranean Veranda Ways project, the Ecosistema Urbano Office Architects has designed and built a bioclimatic boulevard at the New Expansion of Vallecas in Madrid, Spain (Soutullo *et al.*, 2011a; 2011b). The region is depicted in Figure 10.3. The region is experiencing the Mediterranean hot and arid climatic conditions.

a 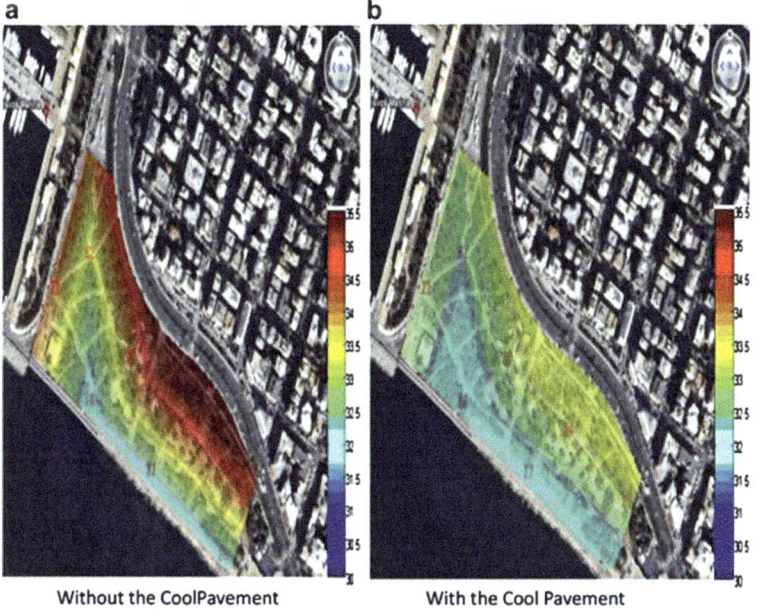 b

Without the CoolPavement   With the Cool Pavement

Figure 10.2
**The Cool Pavements'
Impact on Urban Heat
Island**
Source: Reprinted from
Santamouris, M., Gaitani, N.,
Spanou, A., Saliari, M.,
Giannopoulou, K., and
Vasilakopoulou, K.
Kardomateas, T. (2012).
Using cool paving materials
to improve microclimate of
urban areas – Design
realization and results of the
Flisvos project. *Building and
Environment*, 53, 128–136.
doi:10.1016/j.buildenv.2012.
01.022 with permission from
Elsevier

Figure 10.3 Photo of the ECO Boulevard
Photo by William Veerbeek, Netherlands

Figure 10.4
Map of the ECO Boulevard region
Source: Reprinted from Soutullo, S., Olmedo, R., Sánchez, M. N., & Heras, M. R. (2011). Thermal conditioning for urban outdoor spaces through the use of evaporative wind towers. *Building and Environment*, 46(12), 2520–2528 with permission from Elsevier

NORTH

a *Wind towers & Evaporative systems*

b *Vegetal envelope*

c *Inside projector screen*

The urban space

This avenue is composed of temporary installations that will be reused in other open spaces once the boulevard has developed its own wooded zone, providing a comfortable area for pedestrians. In the urban space three different constructions are built along the boulevard under study as depicted in Figure 10.4. The first one is a set of wind towers and evaporative system, the second one is a vegetable envelope and the third one is a projector screen.

Urban heat island mitigation technologies

As mentioned in the previous section, each cylinder has a different approach aiming to improve the outdoor comfort and increase the social sustainability in the region. The northern one is composed of wind towers with direct evaporative systems installed on the top. The central 'Air Tree' is surrounded by a vegetal envelope that shades the interior area and reduces the ambient temperature thanks to the evapotranspirative process. The third one is covered inside by a screen with a double purpose: to shade the interior of the 'Air Tree' and to be used as a television screen. All the structures are surrounded by sixteen units

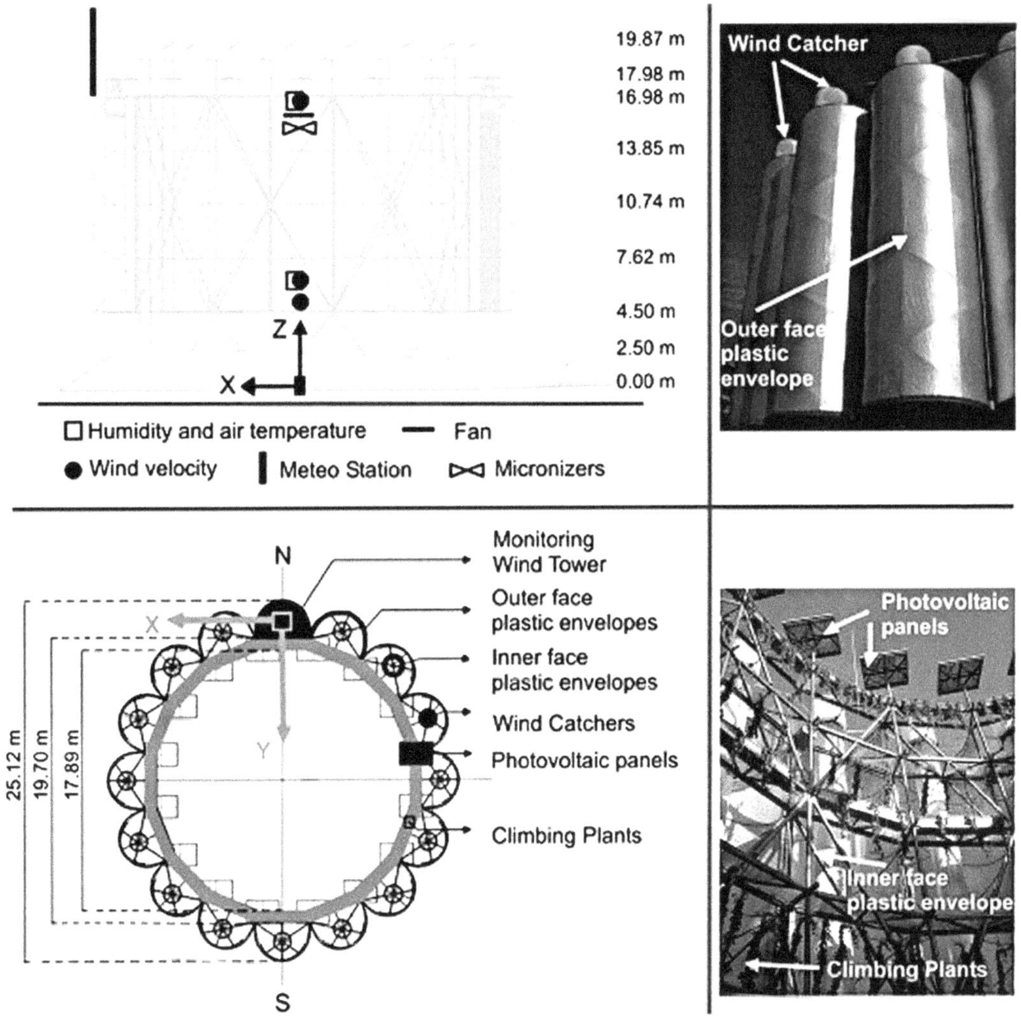
19.87 m
17.98 m
16.98 m
13.85 m
10.74 m
7.62 m
4.50 m
2.50 m
0.00 m
Z
X
Humidity and air temperature
Fan
Wind velocity
Meteo Station
Micronizers
Wind Catcher
Outer face plastic envelope
N
S
X
Y
25.12 m
19.70 m
17.89 m
Monitoring Wind Tower
Outer face plastic envelopes
Inner face plastic envelopes
Wind Catchers
Photovoltaic panels
Climbing Plants
Photovoltaic panels
Inner face plastic envelope
Climbing Plants

Figure 10.5
The evaporative towers in Madrid
Source: Reprinted from Soutullo, S., Olmedo, R., Sánchez, M. N., & Heras, M. R. (2011). Thermal conditioning for urban outdoor spaces through the use of evaporative wind towers. *Building and Environment*, 46(12), 2520–2528 with permission from Elsevier

with four photovoltaic panels each; with a power rate of 165W per module. This photovoltaic installation supplies all the energy requirements and makes the 'Air Trees' independent of the electricity grid (Figure 10.5).

Results

The average temperature decrease due to the evaporative cooling at the pedestrians' height (1 m) is almost 3.5°C with an average saturated cooling efficiency of about 32 per cent. Moreover the Heat index and the thermal sensation Givoni index are calculated based on the outdoor conditions measurements (i.e. air temperature, humidity, solar radiation, wind speed and direction) showing that the pedestrian zones closer to the evaporative tower have increased comfort levels for higher fraction of the day comparing to the regions outside the bioclimatic boulevard.

Tirana, Albania

Locality and climatic conditions

The urban region under study is located in the historic centre of Tirana. The capital of Albania has a typical Mediterranean climate, with hot, dry summers and cool, wet winters. The city suffers from various environmental problems and has experienced a very rapid growth in the construction of new buildings and new urban infrastructures. The speedy urbanism leads to a large leap of population in urban area and amounts of increasing anthropogenic heat that in combination to the use of classical design practices and use of conventional materials like asphalt and dark pavements contributes highly to increase ambient temperatures during the summer period (Fintikakis et al., 2011).

The urban space

The urban space is located in the historical center of Tirana. The city suffers from various environmental problems and has experienced a very rapid growth in the construction of new buildings and new urban infrastructures. The speedy urbanism leads to a large leap of population in urban area and amounts of increasing anthropogenic heat that, in combination with the use of classical design practices and use of conventional materials like asphalt and dark pavements, contributes highly to increase ambient temperatures during the summer period (Figure 10.6).

Urban heat island mitigation technologies

The urban heat mitigation technologies involved measures like the increase of vegetation (planting of trees), the use of shading, as well as the use of materials with appropriate thermal properties. In addition, earth-to-air heat exchangers have been considered to provide cool air in the area. In particular the examined interventions included:

- Extended use of shading and solar control in the area to reduce the surface temperature of the materials and the corresponding heat convection;
- Use of high size trees to enhance shading and evapotranspiration in the considered area;

Figure 10.6
The urban region under rehabilitation in Tirana
Source: Reprinted from Fintikakis, N., et al. (2011). Bioclimatic design of open public spaces in the historic centre of Tirana, Albania. *Sustainable Cities and Society*, 1(1), 54–62 with permission from Elsevier

- Use of light coloured materials to decrease the absorption of solar radiation, decrease the surface temperature of the pavements and streets, diminish the emission of infrared radiation by the materials and reduce the heat transfer by convection between the materials and the ambient air. In particular cool pavements are used in the whole area. The specific solar reflectivity of the selected materials exceeded in all cases the value of 0.65 while the emissivity of the materials was always higher than 0.85;
- Decrease as much as possible the anthropogenic generated heat in the area. This is achieved through the reduction of traffic in the area;
- Maximum possible use of environmental heat sinks and in particular of the ground to offer low temperature fresh air in the area.

Results

The rehabilitation results are evaluated using Computational Fluid Dynamics supported by monitoring the outdoor environmental conditions. The use of passive cooling techniques involving cool materials, solar control and additional vegetation as well as earth to air heat exchangers (Figure 10.7) can reduce the peak summer ambient temperature up to 3 K, while surface temperatures are decreased up to 6–8 K.

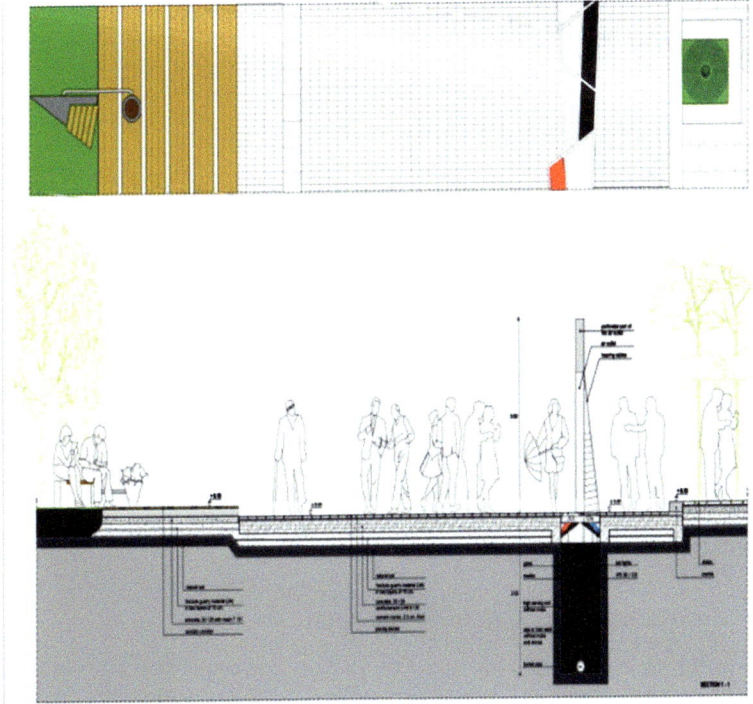

Figure 10.7 Design aspects for the urban rehabilitation
Source: Reprinted from Fintikakis, N., *et al.* (2011). Bioclimatic design of open public spaces in the historic centre of Tirana, Albania. *Sustainable Cities and Society*, 1(1), 54–62 with permission from Elsevier

River Don, Sheffield, UK

Locality and climatic conditions

The area under investigation is situated along the River Don in Sheffield, UK (Hathway and Sharples, 2012). The climatic conditions in the UK are influenced by the Atlantic Ocean and it is generally temperate with cool wet winters and humid summers.

The urban space

The River Don in Sheffield enters the city of Sheffield after passing from the rural area. The urban heat island of the city is evaluated by (Hathway and Sharples, 2011; Lee and Sharples, 2008) and its intensity is almost 2 K. The space under investigation is the one adjacent to the river in locations which are either parallel or perpendicular to it.

Urban heat island mitigation technologies

The main urban heat island mitigation technology is the water of the river and the cooling of the adjacent areas via evapotranspiration. As a result a series of measurements were taken either directly adjacent to the river or running perpendicular to the river bank.

Results

The results show average cooling of nearly 1 K or more for outdoor conditions with air temperature higher than 20°C. Hourly averages of the two hottest periods indicated that the cooling effect depends on the urban form and the materials' used.

Tsim Sha Tsui and Sheung Wa, Hong-Kong

Locality and climatic conditions

Hong Kong is situated in a region with humid sub-tropical climate which is strongly affected by the monsoons. The city experiences hot and humid summers with temperatures higher than 30°C and relative humidity of around 80 per cent.

The urban space

Hong Kong is a densely populated urban region with insufficient buildable land resources. This compact urban morphology can cause thermal heat stress, especially during the hot and humid summer months.

Urban heat island mitigation technologies

Hong Kong's policy from 2000 onwards was the greening of the urban area to provide improvement of is environment. The 1999 Policy Address stated that the government should put power to make Hong Kong a green model for Asian cities. A number of master plans have been elaborated targeting in the increase of greenery in Hong Kong. This is depicted in Figure 10.8. Therefore the main mitigation technology examined is the increase of greenery.

Figure 10.8 The greening of specific regions in Hong Kong
Source: Reprinted from Ng, E., Chen, L., Wang, Y., & Yuan, C. (2012). A study on the cooling effects of greening in a high-density city: An experience from Hong Kong. *Building and Environment*, 47(1), 256–271 with permission from Elsevier

Results

The performance is analysed by measuring the outdoor conditions in the various regions and by numerical modelling using Envi-met 3.1. (Bruse, *et al.*, 2010). The analysis showed a temperature decrease varying from 0.2–1.2 K depending on the size of greenery surface area and height of plantation above ground. It is concluded that a cooling effect of about 1 K is possible when tree coverage is larger than one-third of the total land area when the building coverage ratio is set to 44 per cent, which is the average value in Hong Kong.

University Putra Malaysia, Malaysia

Locality and climatic conditions

Malaysia is a Southeast Asian country situated in a tropical region near to the equator. The country lies between 1° and 7° North latitude and extends from longitude of 100–119° East, with hot and humid climatic conditions and heavy tropical rains. As a tropical country, Malaysia experiences constantly high temperatures and relative humidity, light and variable wind conditions, long hours of sunshine with heavy rainfall and overcast cloud cover through the year. The daily air temperature varies from a low of 24°C up to 38°C, while the recorded minimum temperature is usually during night. Malaysia has high humidity while the mean monthly relative humidity ranging from 70 per cent to 90 per cent all over the year, varying from place to place and from month to month (Makaremi, *et al.*, 2012)

The urban space

The urban space under investigation is the campus of University Putra Malaysia. The analysis is focusing on the outdoor comfort of the pedestrians in shaded spaces under hot and humid conditions. The two study areas depicted in Figure 10.9 are both shaded but with different shading material. The one is shaded by polycarbonate while the second space is shaded by plants and trees.

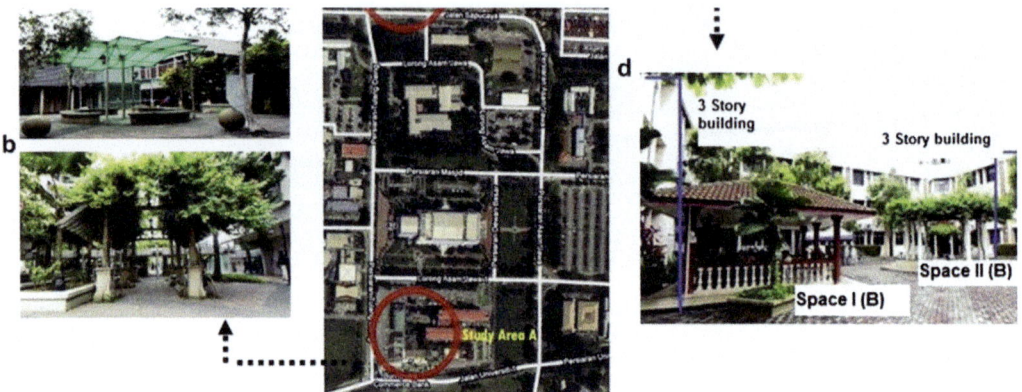

Figure 10.9 The two spaces in the campus of University Putra Malaysia
Source: Reprinted from Makaremi, N., Salleh, E., Jaafar, M. Z., & GhaffarianHoseini, A. (2012). Thermal comfort conditions of shaded outdoor spaces in hot and humid climate of Malaysia. *Building and Environment*, 48(1), 7–14 with permission from Elsevier

Urban heat island mitigation technologies

The urban heat island mitigation technology under study is shading with plants and vegetation.

Results

The outdoor comfort conditions are assessed using the Physiologically Equivalent Temperature (PET) comfort index. The analysis showed that for the overall period the outdoor conditions were outside the comfort range as defined for tropical climate, i.e. PET < 30°C. The spaces shaded by plants had longer periods of thermal comfort.

Green Parking lots, Kobe Japan

Locality and climatic conditions

The specific parking lots which are under examination belong to a public parking space managed by Hyogo Prefecture and are located near Kobe City Centre (Takebayashi and Moriyama, 2009). Kobe has a humid subtropical climate that is mild with no dry season, constantly moist (year-round rainfall). Summers are hot and muggy with thunderstorms. Winters are mild with precipitation from mid-latitude cyclones. Seasonality is moderate.

The urban heat island mitigation effect from changing asphalt-covered parking areas to grass-covered ones was estimated by the analysis of surface temperature decrease and the surface heat budget.

The urban space

The site is located in the residential area of the central ward, Kobe City. The southern side of the site is open, facing the road, with a five-storey building adjacent to the western side and a two-storey building adjacent to the northern side of the site. The size of each parking lot is 5.3 m x 2.5 m, and the access aisle is approximately 7.4 m wide; thus, the size of the site is approximately 56 m x 18 m.

Urban heat island mitigation technologies

The mitigation technology applied is changing the surface from asphalt to grass. Various types of grass has been laid without removal of the older asphalt surface. The urban heat island mitigation effect from changing asphalt-covered parking areas to grass-covered ones was estimated by the analysis of surface temperature decrease and the surface heat budget. In addition, the appearance of the parking lot, the growth of grass, the effects from the weight of a car and the heat radiated from its engine, and the costs of construction and maintenance were also considered.

Results

The urban heat island mitigation effect and the thermal environment improvement effect were estimated by observing the mean surface temperature of the parking lot by using an infrared camera. The tendency for the mean surface temperature of a parking space to decrease with an increase in the green coverage ratio was confirmed, but the mean surface temperature of parking spaces varied greatly depending on the other materials used in addition to grass. From the analysis of the surface heat budget for representative parking spaces, the sensible heat flux was reduced from around 100–150 W m² in the daytime to around 50 W m^{-2} during the night, in comparison with asphalt.

Chapultepec Park, Mexico City

Locality and climatic conditions

Chapultepec Park is a very large park in the western part of Mexico City. Its surface area is about 525 ha.

The urban space

Chapultepec Park includes a number of recreational activities such playgrounds, fountains, a zoo and artificial lakes. Moreover, a cemetery and museums exists inside its area. Surveys have revealed that the condition of trees in the park (many of the same old age) is declining at an alarming rate and air pollution has created canopy openings. In recent years, however, the city authorities have undertaken a park renovation program including the planting of trees.

Urban heat island mitigation technologies

Studies concerning the performance of the park revealed that the condition of trees in the park was declining and air pollution has created canopy openings. A planting of trees and park renovation program was initiated.

Results

The temperature difference between the park area and the nearby urban site in Mexico City are 3–4°C. The smaller thermal inertia of the wooded area results in a higher heating rate at the park during the morning and at midday hours.

Conclusions

All case studies presented here demonstrate the strong impact of cool roofs on the maximum surface temperature which decreases, e.g. by 21.9°C in Florida, 24°C in Italy or 35°C in France. These results highlight the direct potential on the environment temperatures of not cooled buildings. The cool roof technology has also direct consequences on the building thermal loads, which were measured either with the indoor air temperatures or the AC energy demand evolution. Peaks of more than 50 per cent were reached for the cooling energy consumption estimation, being heating penalties lower the cooling gains.

The air conditioning systems are more widely used in hot climates like in Mediterranean regions as Greece, Italy and a large area of United States, concerning the reviewed case studies. The results demonstrate the potential for typical buildings with weak insulation systems. In general variations and savings are directly connected with the building use, climatic conditions, insulation and solar control strategies.

References

Bruse, M., *et al.* (2010). ENVI-met manual. Available online: http://www.envi-met.com/documents/onlinehelpv3/helpindex.htm

Fintikakis, N., Gaitani, N., Santamouris, M., Assimakopoulos, M., Assimakopoulos, D. N., Fintikaki, M., Albanis, G., Papadimitviou, K., Chryssochoides, E., Katopodi, K. and Doumas, P. (2011). Bioclimatic design of open public spaces in the historic centre of Tirana, Albania. *Sustainable Cities and Society*, *1*(1), 54–62.

Hathway, A., and Sharples, S. (2011). Urban river microclimates. In *PLEA 2011 – Architecture and Sustainable Development, Conference Proceedings of the 27th International Conference on Passive and Low Energy Architecture* (pp. 183–187).

Hathway, E. A., and Sharples, S. (2012). The interaction of rivers and urban form in mitigating the Urban Heat Island effect: A UK case study. *Building and Environment*, *58*, 14–22. doi:10.1016/j.buildenv.2012.06.013

Lee, S. E., and Sharples, S. (2008). An analysis of the Urban Heat Island of Sheffield – The impact of a changing climate. In *PLEA 2008 – Towards Zero Energy Building: 25th PLEA International Conference on Passive and Low Energy Architecture, Conference Proceedings*.

Makaremi, N., Salleh, E., Jaafar, M. Z., and Ghaffarian Hoseini, A. (2012). Thermal comfort conditions of shaded outdoor spaces in hot and humid climate of Malaysia. *Building and Environment*, *48*(1), 7–14.

Santamouris, M., Gaitani, N., Spanou, A., Saliari, M., Giannopoulou, K., and Vasilakopoulou, K., Kardomateas, T. (2012). Using cool paving materials to improve microclimate of urban areas-Design realization and results of the Flisvos project. *Building and Environment*, *53*, 128–136. doi:10.1016/j.buildenv.2012.01.022

Soutullo, S., Olmedo, R., Sánchez, M. N., and Heras, M. R. (2011a). Thermal conditioning for urban outdoor spaces through the use of evaporative wind towers. *Building and Environment*, *46*(12), 2520–2528.

Soutullo, S., Sanchez, M. N., Olmedo, R., and Heras, M. R. (2011b). Theoretical model to estimate the thermal performance of an evaporative wind tower placed in an open space. *Renewable Energy*, *36*(11), 3023–3030. doi:10.1016/j.renene.2011.03.035

Takebayashi, H., and Moriyama, M. (2009). Study on the urban heat island mitigation effect achieved by converting to grass-covered parking. *Solar Energy*, *83*(8), 1211–1223. doi:10.1016/j.solener.2009.01.019

INDEX

3 20